T0239266

Pro Database Migration to Azure

Data Modernization for the Enterprise

Kevin Kline
Denis McDowell
Dustin Dorsey
Matt Gordon

Foreword by Bob Ward

Apress®

Pro Database Migration to Azure: Data Modernization for the Enterprise

Kevin Kline
Mount Juliet, TN, USA

Denis McDowell
Raleigh, NC, USA

Dustin Dorsey
Murfreesboro, TN, USA

Matt Gordon
Lexington, KY, USA

ISBN-13 (pbk): 978-1-4842-8229-8
https://doi.org/10.1007/978-1-4842-8230-4

ISBN-13 (electronic): 978-1-4842-8230-4

Managing Director, Apress Media LLC: Welmoed Spahr
Acquisitions Editor: Jonathan Gennick
Development Editor: Laura Berendson
Coordinating Editor: Jill Balzano

Cover image photo by Ferdinand Stöhr on Unsplash

Distributed to the book trade worldwide by Springer Science+Business Media LLC, 1 New York Plaza, Suite 4600, New York, NY 10004. Phone 1-800-SPRINGER, fax (201) 348-4505, e-mail orders-ny@springer-sbm. com, or visit www.springeronline.com. Apress Media, LLC is a California LLC and the sole member (owner) is Springer Science + Business Media Finance Inc (SSBM Finance Inc). SSBM Finance Inc is a **Delaware** corporation.

For information on translations, please e-mail booktranslations@springernature.com; for reprint, paperback, or audio rights, please e-mail bookpermissions@springernature.com.

Apress titles may be purchased in bulk for academic, corporate, or promotional use. eBook versions and licenses are also available for most titles. For more information, reference our Print and eBook Bulk Sales web page at http://www.apress.com/bulk-sales.

Any source code or other supplementary material referenced by the author in this book is available to readers on GitHub.

Printed on acid-free paper

Love is what we have, against age and time and death, against all the powers ranged to push us down. You gave me so much— a new hope, a new heart, a new future — all far better than anything I had before. I dedicate this to you, Rachel, my jewel more precious than rubies.
—Kevin Kline

Dedicated to Antoinette, who has been my cheerleader and biggest supporter, and to Christopher, Bella, Katherine, and Josh, who tolerated geek dad lessons for so many years.
—Denis McDowell

Dedicated to my wife Sarah and our three wonderful kids Zoey, Bennett, and Ellis who put up with me through the countless hours it took to make this a reality.
—Dustin Dorsey

This is dedicated to every member of #sqlfamily that has helped me through the years. As technical professionals, I feel like we have a responsibility to "pay it forward" to help those who come after us just as those who came before us helped us. I am immensely grateful for all who have helped me through the years, and I hope my efforts to pay that forward have proved similarly helpful to others in our community.
—Matt Gordon

Table of Contents

About the Authors

Kevin Kline is a noted database expert and software industry veteran. A long-time Microsoft Data Platform MVP, an AWS Data Community Ambassador, and respected community leader in the database industry, Kevin is a founder and former president of the Professional Association for SQL Server, as well as the author of popular IT books such as *SQL in a Nutshell*. Kevin is a top-rated speaker at industry trade shows worldwide and has a monthly column at *Database Trends and Applications* magazine. He tweets at @kekline.

Denis McDowell has been designing and implementing technology solutions with Microsoft Data Platform technologies for over 25 years. Denis' ten years leading the Application Management practice for a managed services provider and subsequent experience consulting in financial technologies led him to develop broad and deep expertise architecting requirements-driven cloud solutions to meet the business objectives of his customers. Denis is a certified Microsoft Azure Data Platform Engineer and speaks regularly at industry events and conferences around the world. Denis is a consultant at QBE, LLC, a leading management and technology consulting organization for the federal government and defense and intelligence communities, and is currently the Principal Cloud Architect for the US Army's Enterprise Cloud Management Agency (ECMA).

Dustin Dorsey has been architecting and managing SQL Server solutions for healthcare and technology companies for well over a decade. While he has built his career in database administration, he has also spent significant time working in development and business intelligence. During this time, Dustin has gained a keen interest and specialization in cost management around the data platform both on-premises and in the cloud that he has used to save organizations millions of dollars. Dustin is an international speaker and can be seen writing articles on popular SQL websites as well as on his own blog at `http://dustindorsey.com`. He is also active in the community both as a local user group leader and event organizer.

Matt Gordon is a Microsoft Data Platform MVP and has worked with SQL Server since 2000. He is the leader of the Lexington, KY, Data Technology Group and a frequent domestic and international community speaker. He's an IDERA ACE alumnus and Friend of Redgate. His original data professional role was in database development, which quickly evolved into query tuning work that further evolved into being a DBA in the healthcare realm. He has supported several critical systems utilizing SQL Server and managed dozens of 24/7/365 SQL Server implementations. Following several years as a consultant, he is now the Director of Data and Infrastructure for Rev.io, where he is implementing data governance, DevOps, and performance improvements enterprise-wide.

About the Technical Reviewer

Joseph D'Antoni is a Principal Consultant at Denny Cherry and Associates Consulting. He is recognized as a VMWare vExpert and a Microsoft Data Platform MVP and has over 20 years of experience working in both Fortune 500 and smaller firms. He has worked extensively on database platforms and cloud technologies and has specific expertise in performance tuning, infrastructure, and disaster recovery.

Acknowledgments

First, I would like to thank my family who has put up with me and supported me throughout my professional journey as I know it has been challenging at times. A very special thanks to Jim St. Clair and the IT leadership team at Corizon who took a chance on me as a support person and hired me to my first DBA position. That chance set my career down the path it is today, and I will be forever grateful. Thank you to my former managers Justin Steidinger, Jon Buford, Waylon Hatch, and Adam Murphy who challenged and supported me in growth and learning. Thank you to Mark Josephson for being such an amazing person and for all your help, inspiration, and motivation. You played a big role in giving me the confidence to start presenting and becoming more involved in the community that has opened so many doors. Thank you to my fellow coauthors and everyone that helped make this book possible, but a special thanks to Kevin Kline. You not only presented me with the opportunity to write this book, but you have also been a great mentor and friend. Thank you to everyone in the SQL community for all that you do and letting me be a part of you. And lastly, I would like to thank God who has put me in this position and made all this possible.

—Dustin Dorsey

I would like to acknowledge my Dad, Denis Sr., for being the original geek dad and instilling an insatiable curiosity about how things work. Also the technical wizards I have been fortunate to work with and who have helped me in immeasurable ways throughout my career. Scott Brooks and Greg Gonzalez for their laser-focused attention to detail, quality, and excellence. Ken Seitz for his leadership and sponsorship of my pet projects, some of which even worked. Frank Shepherd for continually forcing me to up my engineering game. David Jones and Monty Blight for showing me how leadership should look and taking me to customer visits so I could learn how to talk both to engineers and business stakeholders. Ben DeBow for convincing me that I could make it in the consulting world. Brian Swenson, Keith Lowry, and Michael Lowry for giving me the opportunity to lead a truly transformational effort at the Army. Paul Puckett, Director of the US Army Enterprise Cloud Management Agency for showing me that one can be an agent for transformational change and understand the technical details at the same time. And finally, I would like to thank my coauthors, in particular Kevin Kline, who quite literally started me on my SQL Server journey all those years ago when I saw him speak about performance tuning at a hotel in Charlotte, NC. I left that session hooked and dove head first into SQL Server technologies. The rest is history.

—Denis McDowell

Foreword

More than ever, companies are looking to get an edge. They are looking to modernize their use of technology to respond to events in their industry and across the globe. Whether it is cybersecurity threads or the need to drive innovation at a rapid pace, modernization, especially for data, is critical to maintain or gain market share or just build great products for all of us. Modernize means to do something different, something transformative. One solution to achieve these goals is to migrate your data platform to the cloud. Microsoft Azure provides an ideal destination for cloud migration, especially when it comes to SQL Server.

In this book, the authors give you an end-to-end experience on how to migrate to Azure SQL, a family of cloud services based on SQL Server. Many today look at migration as a very technical exercise to simply move databases and objects from one location to another. This book certainly fulfils those fundamentals but goes further. I like to think of this book as a "setup for success" when it comes to migration to Azure. The authors do such a great job of mixing up practical advice and strategy with technical instructions, resources, and references. For example, this book explores the concept of team dynamics and how it can affect the success of migration. The technical steps of migration are all in this book, but it takes such a more holistic approach including budgeting, cost management, and the use of Azure Hybrid Benefit, a key tool to save money with Azure.

All throughout the book, there is an emphasis on careful planning to save you time, money, and effort. But what I loved about this book is when there is a need to dive deep, the authors are not afraid to go there. Take for example the details they provide when it comes to security and trust which are key concepts to consider for migration. Moving data is a huge part of migration, and the authors don't just show you effective techniques for data migration but give you expertise on how to validate and execute key postmigration steps.

When I look toward a great book, I look at the experience of the authors. Kevin, Matt, Denis, and Dustin have almost 100 years of combined experience in enterprise IT. Between them, they've migrated more than 250 database applications to the Azure data platform. They've built and deployed more than 20 greenfield projects on the Azure data platform in a mix of Azure SQL Database, Azure SQL Managed Instances, and SQL Server on Azure VMs.

To many, migration just means a "shift." This book provides you with a complete experience of making migration a journey toward modernization. Modernization to do something bigger and better. To transform how your business uses data to empower the future. The future you want. The future you need.

Bob Ward

Principal Architect, Microsoft

CHAPTER 1

The Azure SQL Data Platform

As any old timer in the IT industry can tell you, planning *always* helps when managing a project from inception to conclusion. Simple projects can go much more smoothly with an appropriately simple plan, such as a checklist. At the same time, projects that are complex and involve multiple teams are quite likely to need a commensurately more sophisticated plan with the input of many subject matter experts (SMEs) from both the business and IT side of your organization to ensure success.

That is not to say that most projects actually get the level of attention to planning that they need and deserve, considering their importance to the organization. But even simple projects with relatively few steps can often go astray. Many IT practitioners have thought "I know what I'm getting into and I know what I'm doing. This'll be three clicks of the mouse and I'm done," only to find that they now have to spend three days fixing what they thought was going to be a three-hour project.

In fact, many projects go astray even when there is a good plan in place, with the likelihood of difficulties increasing in proportion to the complexity of the IT project at hand. Yet the priority that most managers put on speedy action encourages the IT practitioners to play fast and loose with what should be an orderly, standardized process. Shortcuts are taken. Teams fail to communicate. Bad decisions are implemented with a passing thought to "We'll fix that later." Blind spots are missed. Requirements are misunderstood. Testing and quality assurance steps are shortchanged, with the expectation that the tests will come up roses without the need for any fixes.

The good news is that running mission-critical database workloads in the cloud is no longer a brave, new computing paradigm. It is a tried and true approach to enterprise computing with many advantages to other methods of operating your corporate information technology. But just because the cloud has been around long enough and grown to be popular enough to wear off the jagged surfaces of the bleeding edge, it is

© Kevin Kline, Denis McDowell, Dustin Dorsey, Matt Gordon 2022
K. Kline et al., *Pro Database Migration to Azure*, https://doi.org/10.1007/978-1-4842-8230-4_1

not a cakewalk, especially for those of us who have spent our careers focused on locally operated data centers. That's where this book comes in.

This book is for those who know the importance of their data assets, can envision the difficulties that they might encounter moving them to the cloud, and have the wisdom and foresight to learn as many lessons, both the failures and successes, as they can from those who have gone before and successfully migrated their on-premises databases to the Microsoft Azure cloud.

In subsequent chapters of the book, we will share best practices for the most important steps in a database migration project, covering everything from an on-premises environment to the Microsoft Azure cloud.

Azure Core Services and Concepts

Broadly speaking, Microsoft Azure provides hundreds of distinct services that enable you to build and manage powerful applications using Microsoft Azure cloud services. Covering all of those services and feature offerings is far beyond the scope of this book. However, we will take a few minutes here to make sure to provide you with an overview of key Azure services and concepts.

First, Azure provides four broad categories of services: Infrastructure, Platform, Security and Management, and Hybrid Cloud. Each of these categories contains many more point services and products. For example, the Security and Management category contains one product that you'll use on a daily basis, the Azure Portal.

Among these many individual services, we have the *core services*, which include *Compute, Network, Storage,* and *Database.* Compute is the bedrock foundation of Azure and includes services like Azure Virtual Machines, Azure Container Instances, Azure Kubernetes Service, and Azure Virtual Desktop. Azure networking services include Virtual Networks, VPN Gateways, Azure ExpressRoute, Azure Load Balancer, and more. Storage services include Azure Blob Storage, Azure Disk Storage, Azure File Storage, and other services. Finally, and perhaps most importantly for this book, we have a variety of data services like Azure Cosmos DB, Azure SQL, Azure SQL Managed Instance, Azure Database for MySQL, Azure Database for PostgreSQL, and Azure's big data and analysis services.

Conceptually, there are a few important details you should learn about building and deploying applications to Azure, as shown in Figure 1-1. When you build an application in the Azure cloud, you'll provision the services and features you want by first creating an Azure account, that is, your global unique credentials used to sign on Azure and access your subscriptions. (Same people refer to this as a *tenant*). An account/tenant is associated with a single entity, such as a person, company, or organization. You are also likely to manage your Azure account using Azure Active Directory (AAD) for authentication and authorization, providing granular permissions to the services you and others in your organization are using in the cloud.

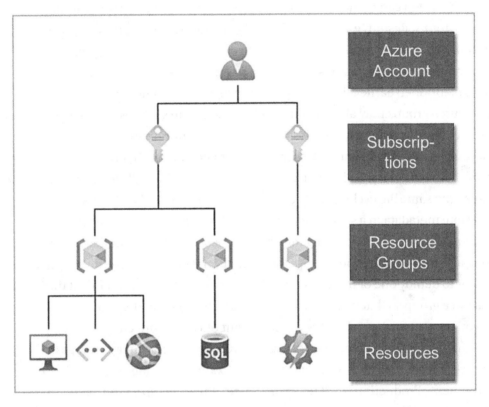

Figure 1-1. *Azure provisioning of services built atop Azure Core Services*

As we see in Figure 1-1, the hierarchy within Azure enforces a one-to-many relationship. One tenant/account can have multiple subscriptions, but subscriptions belong to only one tenant/account. A subscription can have many resource groups, but a resource group belongs to only one subscription. And so on.

Following the creation of your enterprise's Azure *account*, you can create one or more *subscriptions* for billing purposes and to provide separation and control when managing how you spend your budget in the cloud. For example, you might create separate subscriptions for your Dev and QA departments, with additional subscriptions to support billing directly to the budgets of your organization's business units. You may also want to specify one or more Azure policies at the subscription level, for example, to send the Azure administrator a notification that a specific budget quota is close to reaching its limit. You can also associate one or more Azure subscriptions so that they trust your AAD tenant.

Most organizations have multiple subscriptions, but not all. Within your subscriptions, you can manage specific resources by defining and provisioning *Azure Resource Groups*. Resource groups are logical collections of resources belonging to the same application stack, environment, and life cycle. For example, a typical resource group for the purpose of database migration to Azure would include the Azure core services mentioned earlier, like database, storage, network, and compute resources. But your resource groups may also include many other Azure services so that you can access a variety of other specific capabilities when you build your Azure solution.

Resource groups from one subscription are isolated from other resources in other subscriptions. It's good practice to use a resource group to logically aggregate resources that share the same life cycle for easy deployment, updates, and deletions. Resource groups store metadata in a specific location, which can be very important if you must meet compliance regulations concerning the physical location where your data is stored. After creating a resource group, you can use the Azure Portal or Azure Resource Manager (ARM) templates to deploy one or more resources to the resource group. You might then also *lock* the resource group so that no critical resource can be altered or deleted.

If you want to further refine the management of your Azure environment with as many as 50 *tags,* composed of a name/value pair, you can apply tags to a resource, a resource group, or a subscription. Examples for tags might include "Environment=Production" or "Owner=Finance". Tags have size limits and are case-sensitive. Microsoft recommends managing tags in a hierarchy as described at Tag resources, resource groups, and subscriptions for logical organization - Azure Resource Manager | Microsoft Docs.

Within this construct, you can also assign *Azure role-based access control* (Azure RBAC) to one or more users, groups, service principals, or managed identities. Azure RBAC provides a wide selection of built-in roles, like *Owner* or *Contributor,* so that you can effectively manage access to the Azure resources in your subscriptions. These built-in roles are frequently all you will need for proper user authorization, but you can also create your own Azure *custom roles* to further refine resource authorization. A full discussion of these concepts is beyond the scope of this book, but check online for a full discussion at Azure RBAC documentation | Microsoft Docs.

Finally, this section would not be complete without a discussion around some basic concepts of designing your Azure architecture to support high availability and disaster recovery as required for your environment. An in-depth discussion of designing Azure infrastructure for high availability is beyond the scope of this book, but we did want to make sure some basic concepts and terms were covered in our introduction as they can, if needed, be a part of the Azure architecture you deploy to support the migrations discussed in depth in the rest of this book.

Understanding what an Azure region is will be fundamental to any deployment decisions you make going forward. Azure regions are data centers deployed, as Microsoft states it, "within a latency-defined perimeter." They are connected to each other via dedicated networks. That means that while a region may be a singular place in our minds, it can be comprised of multiple data centers. This leads us directly into the concept of availability zones in Azure. Azure availability zones are separate locations within each Azure region – meaning that local failures in a single data center should not impact the other physically separate locations within the availability zone. Microsoft guarantees that a minimum of three separate availability zones will be present in all availability zone–enabled regions. If a service you are provisioning offers you zone redundancy, that means it is taking advantage of the availability zones on offer in the region you've chosen for deployment.

An additional concept that is useful to understand is the concept of paired regions in Azure. While not all services support this cross-region replication, it is another layer to consider as you design your deployment for high availability. For services that support this concept, it means those services will be supported not just within the availability zones in a region but outside the region with another region as well. It further minimizes the possibility of a local (or regional) event impacting the availability of your Azure-deployed services.

In layman's terms, these concepts are important to understand because taking advantage of them can go a long way toward eliminating single points of failure in your deployed resources. Major cloud providers have designed amazing redundancy and resiliency into their systems, but we cannot rely solely on that when deploying resources. As always, understanding the trade-off between high availability, the cost to support it, and the need for it is critical. If that is a consideration for your project, understanding the concepts discussed in the last three paragraphs will be helpful for you.

Migration and Modernization

There are many drivers and motivations that push organizations to undertake a migration to the cloud. Sometimes, organizations might be motivated by strong internal considerations relating to their technology infrastructure, like

- Simplifying their business by no longer running their own data centers.

- Getting extended end-of-life support for important products. For example, SQL Server 2008 on-premises is no longer supported by Microsoft, except if it is housed in an Azure VM where it has extended support.

- Access to highly scalable infrastructure in cloud data centers so that they can provide much better support to requirements for a highly elastic application design.

- Easier management and IT operations for hybrid environments.

Other business-centric motivators often come into play with migrations to the cloud, like

- Ease of operations when expanding operations in new markets, particularly when those markets require data residency, that is, that data collected for their citizens is stored in data centers within their national jurisdiction. For example, Germany requires that international organizations that collect data on German citizens must process and store that data in Germany.

- Heightened security and compliance capabilities, such as those offered by Microsoft Azure, which far exceed those common to many self-run data centers.

- Business functionality improvements by taking advantage of cloud-native application designs and other developer and application innovations, like microservices and containers.

These are excellent reasons to move to the Azure cloud. We'll also spend some time later talking about failed migration projects and how they went wrong in an upcoming section entitled "Cautionary Tales." But first, let's take a quick look at the most common high-level migration strategies.

Migration and Modernization Strategies Made Simple

Jargon is always a big impediment for those who are diving into a new arena of knowledge, particularly in vocational and professional situations. Cloud migration and modernization are no different.

Before going too far, let's cover some of the terminology describing migration strategies at their highest level, since we will refer back to these terms frequently throughout the book. Let's take a look at Table 1-1 to see a quick description of the most prominent migration strategies.

Table 1-1. *Terms for cloud migration scenarios*

Term	Description
Rebuild	Take the existing stack and convert it to one of the Azure PaaS options, such as Azure SQL Database or Azure SQL Managed Instance. This strategy may represent a high level of new code, database designs, and application architectures to take advantage of the new capabilities of the PaaS environment.
Replace	Take an existing stack from your data center and convert it to a wholly SaaS solution. This is the most time-consuming and difficult strategy, since it involves an entire rewrite of the application, supporting services, and possibly database design and database programmable modules, like stored procedures and user-defined functions. When done well, the replace strategy (also known as "cloud native") can offer the greatest performance improvements and new features, but it also includes the greatest risk.
Rehost	Take an existing stack from your data center, when it is managed entirely as Virtual Machines (VMs), and then move them to an identical configuration of Azure VMs. Also known as "lift and shift" since this strategy represents no new code or design work, only new hosting servers.
Refactor	Take the on-premises stack, and then rewrite many of its components to take advantage of and optimize for the Azure cloud. This strategy represents a moderate level of new code, as well as new database and application designs. This is also sometimes called "rearchitect."

In addition to the common terms used to describe the high-level strategy employed to complete your cloud migration shown in Table 1-1, there are also a variety of important terms used to describe the types of cloud architecture we intend to use, as shown in Table 1-2.

Table 1-2. *Important terms for cloud architectures*

Term	Description
IaaS	Infrastructure as a Service (IaaS) is a type of cloud computing service that offers essential compute servers, storage, and networking resources on demand using Azure VMs. IaaS frees you from buying and managing physical servers. But you must still purchase, install, configure, patch, and manage your own software – including operating systems, middleware, databases, and applications. Synonymous with the term "lift and shift."
PaaS	Platform as a Service (PaaS) is a superset of IaaS. It includes the compute servers, storage, and networking of IaaS. It also includes middleware, development tools, business intelligence (BI) services, databases, and more. For an Azure migration project, PaaS refers to systems built on Azure SQL Database and Azure SQL Managed Instance.
SaaS	Software as a Service (SaaS) gives you a complete cloud-centric software solution that you purchase on a pay-as-you-go basis from Microsoft Azure. Examples of SaaS products include email, travel booking applications, customer relationship management (CRM) applications, and ride-hailing services. Azure takes care of all of the underlying infrastructure, middleware, application software, and application data that are running in the cloud, as well as ensures the availability of the app, and secures the app and your data. SaaS applications are usually brand new, called *greenfield applications*, although we have seen some customers use Azure Service Fabric to rebuild older, monolithic applications (usually based on IIS/ASP.NET) into an elegant, cloud-based application using multiple, manageable microservices using domain-driven design principles.
Stack	The infrastructure + data and databases + application under consideration.
Landing Zone	A landing zone is your initial Azure environment for hosting the stack you plan to migrate, preprovisioned according to your requirements and strategy. Your landing zone will typically undergo many changes, expansions, and adjustments before it is ready to support a full production workload. More details at What is an Azure landing zone? - Cloud Adoption Framework	Microsoft Docs.

For a more comprehensive review of all potential terms and concepts, as well as recommendations for building a large, enterprise-scale Azure infrastructure, read more at `https://docs.microsoft.com/en-us/azure/cloud-adoption-framework/ready/enterprise-scale/architecture`.

The Five Disciplines of Cloud Governance

The Microsoft Cloud Adoption Framework provides a comprehensive set of business and technology guidance to better organize and align your business and technical strategy, provide best practices, and offer tools to help you achieve a successful migration. Investing in strong cloud governance is worth every penny. Not only does a cloud governance framework help you envision business risks, it also helps you create well-considered policies and processes to support with and align to the needs of your organization. (Read more of the Cloud Adoption Framework at `https://docs.microsoft.com/en-us/azure/cloud-adoption-framework/`.)

When discussing a cloud migration project, it is helpful to consider important aspects of the project outside of technology alone. As pointed out in the Cloud Adoption Framework, there are five major pillars or disciplines of governance when migrating to the cloud.

The five disciplines of cloud governance are as follows:

1. **Cost management**, in which organizations anticipate and monitor their costs in the cloud and, where needed, limit the total amount spent. When needed, the organization can also scale up resources to meet demand, in line with additional costs. In addition, cost management governance dictates that accountability is extremely important, so the migration team should also create cost accountability policies and processes and authorize staff. Chapters 3 and 4 go into cost management in great detail.

2. **Security baseline**, where the organization ensures compliance with IT security requirements by applying a baseline of security policies and technologies to all migration efforts. Chapter 8 discusses elements of building an Azure security baseline.

3. **Resource consistency**, which ensures consistent application of resource configurations for better enforcement of policies around on-boarding new migration projects, backup and recovery, and discoverability for Azure-based resources. Chapter 6 goes into the details about ensuring you have enough resources for your Azure cloud landing zone.

4. **Identity baseline**, whereby our organization ensures a baseline for identity management and accessibility to the Azure environment through consistently applied and enforced role definitions and assignments. Chapters 7 and 8 cover some elements of identity management applicable to an identity baseline.

5. **Deployment acceleration**, offering a centralized, consistent, and standardized migration and postmigration experience through the use of deployment templates, such as those provided by Azure Resource Manager (ARM) templates. Chapters 9 through 11 cover a variety of best practices that accelerate your Azure deployment.

The Cloud Adoption Framework and its operating model for governance is an expansive set of documentation with many valuable recommendations, guidelines, and best practices. We recommend that you study this information as soon as convenient or, at the latest, after your first migration pilot project but before you have many applications running in Azure. Read more at The Five Disciplines of Cloud Governance - Cloud Adoption Framework | Microsoft Docs.

Cautionary Tales

A smart person learns from their own mistakes, but the wise person learns from the mistakes of others. In the interest of wisdom, we want to make sure you are aware of the many reasons why database migration projects fail. Throughout the rest of this chapter, we will tell you a few cautionary tales to help you know what to avoid.

When the C-Suite Falls for the Hype

Anyone who has spent more than a few years working in an IT career will tell you that the biggest stumbling blocks are frequently organizational or interpersonal aspects of a project instead of the technology itself. In fact, some of our more cynical colleagues might even say that people are an even bigger source of issues and obstacles than technology itself. At least with technology, you do not have to soothe its ego.

Ego may rear its ugly head at any level of an IT organization, as you know if you have ever witnessed two or more IT managers argue over turf or defend their silos. But one especially pernicious source of problems arises in the C-suite. It is an all-too-common situation for a CIO or other executive to attend a big, glitzy conference with a flashy keynote delivered by the CEO of some tech giant or another. And those keynotes, while not untrue, are a form of marketing that attempt to put new and often untested technologies into the best possible light. After returning from an event like that, it's not uncommon to see our executive leaders have stars in their eyes and they have fully bought into a brand-new technology's promise of "saves money!" and "accelerates delivery!" when in fact the technology is so new that its early adopters are more realistically beta testers who are paying for the privilege of finding bugs in the new, bleeding edge technology.

When the CIO returns from a trip like that and swoops into the office of their managers, those managers know they are in for a headache. Many front-line IT managers know that we do not face strictly technological challenges; rather, we face *business challenges that we chose to solve using technology*. Any time we want to make a major change to our fundamental IT systems, we open up our systems to risk. Not only that, major changes to our IT systems are predicated on our people and their talents and experiences.

So when the CIO promptly calls a meeting with the database team(s) and declares "We are going to save so much money; we are moving to the cloud!" they are not thinking about the difficulties in store with the migration or with the lack of skills and experience with the new technology. Most likely, the CIO has ego on their mind, perhaps by making a major mark on the enterprise or by saving money in ways that their predecessors never did.

How do you think this story will end? Badly, no doubt.

Whenever we see knee-jerk initiatives that are not solidly based on business needs nor grounded in measurable outcomes, troubles are in store. It may be possible to pull off a successful migration, but probably only through a heroic and exhausting expenditure of energy by the staff. After all, it's very difficult to know when you've been successful if you don't really know what success looks like.

When Cost Is the Only Consideration

Although there are other appeals that win over business executives, the most frequent (and usually inaccurate) appeal is that migrating to the cloud will save money for the enterprise. This pitch is so appealing precisely because IT is viewed from the old twentieth-century point of view that IT is only a giant cost center within the enterprise.

When the CIO has many years of buying software licenses and hardware in multimillion-dollar acquisitions, this method of purchasing becomes the norm for them. But they are in a situation where their many years of experience leave them unprepared for what it costs to actually migrate an application to the cloud or how to keep cloud costs in check once it is implemented.

Now that we, as an industry, have had time to see how budgets play out for the early adopters, we have learned that migrating the IT show of an enterprise from on-premises to Azure rarely saves money. That's not to say that it will never save money. It is certainly possible, especially with careful and conservative planning. But it is much more realistic to recognize that the major financial benefit is not direct cost savings. Instead, its biggest financial benefit is simply to shift IT costs from the capital expense side of the balance sheet over to the operational expense side of the general ledger. (Cloud has many other benefits of great value, such as scalability and flexibility. However, it requires a lot of planning and preparation to reap any financial rewards). Where before, multimillion-dollar acquisitions could only be amortized over many years, now those expenses can be immediately accounted for on a monthly basis.

Even after learning that fact, many CIOs move forward not realizing that they now need to implement an entirely new way of budgeting through monthly billing and probably through bill-back procedures. In addition, when the CIO is used to seeing very large purchase orders every few years, they can be in for a very big surprise when they start to receive moderately large POs every month.

Monthly bills can be quite stressful and may introduce political conflicts when the CIO seeks to distribute those monthly costs to other departments that incur them. Arguments and turf wars can increase. And instead of delivering on the promise of a better and more capable IT infrastructure, many such projects either underdeliver on their initial promise or even fail outright.

When Middle Managers Are Set Adrift

After the decision to implement a major paradigm shift like migrating from on-premises to the cloud, veteran IT managers know that it's now time to negotiate a variety of requirements. All previous assumptions about requirements should come off the table, from the dates and deadlines of the project, to the functional and business requirements of the project, to disaster recovery and high availability, to security and compliance, to detailed specifications for adequate performance. Unfortunately, veteran IT managers are in short supply. And, by definition, veterans are experienced with how to fight the last war, not necessarily the next one.

Vague Is As Vague Does

When executive leadership is ambiguous or vague, the task of delivering on an Azure migration project falls to the middle managers and IT operational staff. An extensive body of academic research and corporate analysis accumulated over the past several years help us to understand where cloud migration projects run aground when lacking strong executive vision and leadership.

What happens when a veteran is given ambiguous directives outside of their comfort zone? Try to make the project as comfortable as possible, naturally.

Add to that the fact that almost all IT managers are overloaded and suffer under the "tyranny of the urgent." This leaves most IT managers feeling compelled to choose the easiest option, in the absence of a strong executive leader who wants an optimal outcome, not simply an adequate outcome. This often leads directly to a migration solution that meets the requirements of the IT manager(s), but not the end users. In situations like this, it's not uncommon to see the migration project essentially replicate the on-premises data center infrastructure in the cloud without much, or any, attention paid to end-user metrics of success, like latency, responsiveness, or availability.

Fear: The Ultimate Motivator

In times of deep change, a normal person responds at an emotional level with fear and uncertainty. That's perfectly normal. However, fear of change and the responses to that fear by IT managers often have devastating consequences for important migrations projects.

First, IT managers may fear losing control of their silo or fief – to consultants, to shadow IT, or to a new crop of staff that have entirely different skills and mindset. Without a strong executive leader to soothe those fears, IT managers often resist the migration, add friction to the project, or simply fail to provide aid to those needing it.

Second, IT managers may fear that their years of experience and store of experience make them less valuable. In this situation, we often see that fear manifests into choices for cloud services or providers that the IT manager is comfortable with but who add no value to a cloud migration, at best, and actually harm the migration project, at worst.

Let's explore this dynamic deeper. As found in a 2018 survey of 550 executives conducted by IDG, more than one-third (38%) report their IT departments are under pressure to migrate all applications and infrastructure to the cloud. That's right – ALL applications and infrastructure.

By now, we have seen that the cloud is just as secure as your data center, probably more so. Having said that, most organizations have a very poor record on their own for properly addressing security, compliance, and privacy concerns in their previous IT projects. The cloud can help a lot in this respect, especially when, according to experts, security operations are included in the cloud migration process from the get-go, ensuring that those plans adequately incorporate all requirements.

Lessons from Market Research

Upon further study of the survey data, we see that about 60% of the respondents recognize that their most important challenges for the migration project are political and interpersonal. They know that success for their project requires strong executive buy-in and leadership.

Sadly, the survey results show that the same 60% of IT managers in this situation take one or more easier paths. Examples here include requesting executive support only for things like purchasing better tools or, even worse, avoiding any of the difficult strategic questions by settling for retrofitting the current on-premises infrastructure as a direct "lift and shift" to the cloud. One example of this is implementing VMWare Cloud on Azure, which can offer the benefit of a fast and low-risk cloud migration, but at a very high financial cost. Many IT managers think that because they are experienced with a specific virtualization vendor's platform, they should simply replicate that same infrastructure in the cloud without making changes. Quick and easy, right?

Ironically, no. The same surveys cited earlier report that more than 65% of enterprise managers know that they have significant work ahead of them especially in terms of integrating their cloud databases and services into their core operations and applications. This poses a significant enough problem that most enterprises expect that hybrid computing arrangements will be the norm going forward, rather than a simple process of "lifting and shifting" to the cloud.

By sticking with the old, staid technologies they have used for decades, IT managers basically ensure that their migration project reaps the worst of both worlds of cloud and on-premises, not the best. For example, you still must do OS patching, install security hotfixes, and provision servers when using the typical "lift-and-shift" strategy. Yes, rehosting may mean that the migration moves quickly. However, the cloud target environment is oftentimes different enough that even speed of execution is not an advantage.

The Benefits of Cloud Computing

What are the primary benefits of moving to the cloud, if not money or a quick rehosting of the IT estate into a public cloud as discussed in the section immediately prior? Again, we can turn to market intelligence to tell us the five big benefits of cloud computing:

1. **Increase efficiency**, in which IT staff spend more time adding value to IT and business operations and less time simply keeping the IT ship afloat. For example, cabling and racking servers is no longer your problem.

2. **Security**, because the Azure cloud is already more secure than the average data center, data is automatically encrypted at rest and is constantly kept updated with the latest security fixes and findings.

3. **Data storage**, due to advantages of scale and readiness for near-instantaneous provisioning. A couple clicks on the Azure Portal can provide additional and/or faster storage.

4. **Flexibility**, because new services and capabilities can be enabled at the click of a button with no delays for licensing, processing purchase orders, or installation.

5. **Scalability**, both by moving workloads to more powerful compute tiers and back down, as needed, and deploying intrinsically elastic service tiers like Serverless computing.

As you will read elsewhere in the book, you can amplify all five of these benefits by taking a methodical approach when you migrate to the cloud. Rehosting rarely provides these benefits.

Repatriation

Another relevant finding from industry market research is the concept of *repatriation*. This is a fancy word for "we failed at our cloud migration and now have to move back on-premises." In one survey from 2020, 80% of the survey respondents repatriated workloads from a public cloud provider in the last two years. In another study, 74% of respondents had already repatriated workloads from the cloud.

Why are so many cloud migrations failing the test of time? The top drivers in both cases, although not necessarily in the same ranked order, are cost, control, performance, and security. To explain:

- **Cost:** Many enterprises overprovision when they buy servers for their on-premises data centers. Conversely, when migrating to the cloud, most IT managers choose a lower cost service tier, if not the lowest cost. In this context, after migrating, the IT managers end up with a monthly dose of sticker shock because the cloud bill still costs more than they expected but doesn't deliver the performance they assumed. Never mind the fact that they didn't learn about ways to control costs, like auto-provisioning/deprovisioning, elastic pools, and other ways to keep costs low. Also remember that many chief financial officers (CFOs) are excited to see IT costs move from the capital expenditure side of the balance sheet to the operational side. This means that the CFO can write off the entire balance of IT spending each month, rather than using the lengthy write-offs of the amortization schedule used on capital expenditure. However, if the CFO sees monthly expenses for cloud computing go sky high, there is typically a swift and unhappy response.

- **Control:** This aspect of cloud computing usually surfaces in organizations where there is decentralized leadership. In on-premises data centers, we see issues of control for example when every developer has SA privileges on their SQL Server database server. That's out of control, clearly, but extend that into the cloud where each person with sysadmin-level privileges can enable a new feature costing thousands of dollars per month and chaos can quickly ensue. Some enterprise IT shops are infamous for making changes in production. When this approach is taken to the cloud, disaster is practically assured.

- **Performance:** As you'll see in Chapter 2, performance benchmarking is a very important part of migration planning. You'll need to make sure that users are not dealing with latency for the new cloud application that they didn't have with the on-premises version. Otherwise, you'll face a very quick outcry from the user base. But if you haven't already benchmarked on-premises performance, how can you be certain that your landing zone in the cloud will be able to support your users' needs? You can't. And after you've migrated without a performance benchmark, you're never going to be fully certain when performance in the cloud is abnormal because you didn't take the time to fully quantify what is normal performance.

- **Security:** Ironically, the Azure cloud is far more secure than the average enterprise IT organization by default and light years ahead of below-average organizations. The problem here is that when enterprises have a long history of apathy or complacency about security, they tend to export that posture to a public facing cloud. "Public" being the key word here. Now, their IT infrastructure is exposed to all the hackers. Even a highly capable default level of security cannot fix a stupid approach to security. Furthering security as a motivator for repatriation, a host of new legal compliance regulations around the world add new, sharp teeth to an area of IT compliance that was formerly characterized by a legal slap on the wrist. In some cases, early adopters of cloud computing have discovered that they must move their data either on-premises or to another data center in order to comply with national data residency requirements, such as those found in Germany. When IT organizations migrate to the cloud with little analysis, they often discover that their legal risk is so great that it is better to not reside in the cloud at all than it is to reside in the cloud without a proper and well-tuned security shield.

The number one reason, in both cases, is failing to plan for the level of performance needed by the workload and the balance between performance and cost. If you don't plan in advance for what you need in terms of performance, yes, you can scale up your cloud computing investment and simply buy more resources. For example, storage performance can be upgraded from slow hard disks to speedy SSDs. But is the performance you need at a price point that you can afford? Many times it is not.

Pilot Projects: Which to Choose or Avoid

In some cases, we see the primary reason for repatriation is a poor choice of pilot project – or no pilot project at all.

Any time an enterprise undertakes a major new technology shift, it is strongly advised that the project team wisely choose their pilot project, or simply "the pilot." We have seen these patterns emerge during the shift from mainframe computing to PC computing, then from PC computing to Client-Server computing, then from Client-Server to Web 1.0 and Web 2.0 computing. At each inflection point, the smart enterprises begin their shift from old to new technologies using one or more pilots.

Tip   Remember to plan your migration as a series of iterations, refining your design at each iteration. You may have to make changes to the Azure landing zone, to the stack you are migrating, or any number of other elements of the plan. Gradually improve your migration over time rather than striving for a perfect migration at the outset.

The ideal pilot will be large enough in scope that it could potentially yield a substantial improvement for the enterprise, but not large enough to expose the enterprise to major risks. Other key characteristics of an ideal pilot usually balance against two polarities, as shown in the following bullet list:

- **Champions:** As mentioned earlier, the single biggest determinant to a successful migration project, indeed probably *any* project, is an empowered executive champion with a strong vision for the project. Shortchange your migration of this characteristic only if you want to see your project struggle. In addition, your executive sponsor provides one often forgotten advantage – having an executive who promotes the success and value of your migration team and your pilot project to other executives.

- **Duration:** Ideally, the pilot project is long enough in duration that the team learns solid lessons about the new technology, but short enough that executive and operational leadership can assess their likelihood of success in a reasonable timeframe. Perhaps a few months is optimal, since that gives team members enough time to enjoy the benefits of the migration but not so large or complex that it becomes easy to get bogged down. In addition, Microsoft recommends that the migration project should be broken down into two-week sprints, culminating in an analysis of lessons learning into the next sprint.

- **Criticality:** When choosing a pilot, you want one that offers big improvements to the enterprise, but that will not endanger anyone's career should it fail or encounter a lot of setbacks. If too much is riding on the pilot, pressure on the team of implementers multiplies perhaps to the point of sinking the pilot, especially if there are delays or difficulties. (Note: There are *always* delays and difficulties.) You also don't want a pilot that is so inconsequential for the enterprise that there isn't a measurable business improvement or that critics can claim things like "anyone using any technology could've done as well."

- **Size:** When choosing a pilot, it is important to calculate the "size" of the project by both the number of staff assigned to the project and the number and complexity of requirements the pilot project seeks to solve. Ideally, a single team unit of four to six teammates is ideal for the pilot, since larger size teams face increasing difficulty communicating and prioritizing. On the other hand, the pilot needs a well-defined set of requirements that cover functionality, performance, security, availability, recoverability, and maintainability. Once the pilot project is successful, future such migration projects will likely use the development artifacts of the pilot (such as requirements documentation) as their original template. Do it right from the trailblazing pilot and all subsequent migration projects have a higher probability of success.

The pilot project should also focus on a well-defined set of helpful end users who are patient with mistakes, communicate freely, and make themselves available for discussion and prototype reviews.

Additional research has shown one specific type of pilot to be especially risky – those with a multitude of poorly documented data pipelines. For example, one migration project we have seen involved an application that started life in the early 1980s on the mainframe, which received automated daily data feeds from the field. Later, client-server components were added, extracting data from the mainframe to a staging SQL Server using SSIS (SQL Server Integration Services). The data was munged and transformed on the staging server and then exported to a production SQL Server database where another application added additional data through an internal web application. A couple years later, the production database was augmented with a SQL Server Analysis Services (SSAS) data warehouse, which in turn was accessed by a variety of analysts using tools as diverse as Microsoft Excel, Tableau, and Power BI to access and analyze their data for specific situations.

Although the migration team for this enterprise valiantly worked much overtime to successfully replicate their old-school application in the cloud, they kept encountering other users who they'd never heard of before. Each new group of users had their own uses for the data, their own workflows, and their own political champions. With every passing week, the migration team faced more and more friction from other teams until the project became an unhappy quagmire. Eventually, the pilot project was dramatically rescoped to migrate only one small part of the entire on-premises application, which is still running to this day.

The Skyscraper with No Architect

Have you ever heard of a skyscraper that had no senior architect? No – because a skyscraper without an architect won't rise and stay upright for long. Conversely, what about massive buildings with a lot of architects? Those happen quite frequently, and usually, they're a mess. These monumental buildings might stay erect, but they usually display one or more incongruencies, an unappealing aesthetic, and a lot of difficulties with maintenance and upkeep.

When your organization has skilled and experienced IT architects, you'll see direct and clear benefits. Architects are equally important when setting out to build important parts of the enterprise IT organization, especially when the technological underpinnings are new and unfamiliar to the rest of the IT staff.

IT as a discipline has many types of architects, but your pilot project for cloud migration should have a skilled and experienced *enterprise architect*. There's no doubt that it'd also benefit from database architects and application architects, who can each add value to the quality of their components in the migration. However, the enterprise architect is specifically trained to take in the entire scope of requirements of the pilot, assess the pros and cons of a variety of alternatives, and make choices that best fulfill the requirements of the pilot.

In fact, good enterprise architects (EAs) can quickly review your project requirements and identify whether those requirements can sufficiently accomplish the project. And remember, *someone* must fill the role of EA even if no one on the team has the right skills.

Worth Every Penny

Consequently, we strongly advise that you either hire an experienced EA, if you don't already have one on staff, or commit to providing adequate time for training and the training resources to equip a current member of your team to fulfill this role. Since talented EAs are both rare and expensive, it is most likely that a current team member will take on this role, perhaps from your application or database architects, or perhaps from your current project managers. In any case, expect painful "opportunities" for lessons learned during the migration as your novice EAs develop their skills.

What will be expected of your EA throughout your pilot?

1. **Discover and document:** The EA acquires and/or creates documentation that provides a thorough understanding of your current on-premises data estate and infrastructure.

2. **Assess and envision:** The EA assesses the current data estate, infrastructure, and application(s)to be migrated and, in response, develops a plan for a migration "landing zone" or target in the cloud to serve as the migration's destination.

3. **Design and pilot:** Technology always offers more than one solution to a given business problem, sometimes a lot more than one solution. The EA compares requirements of the pilot to options for the technical implementation of the project and returns to the team with one or more recommendations, including the fundamental strategy for the migration. Most organizations choose from four or five strategies.

4. **Rehost:** This is the simplest strategy, in which on-premises servers are virtualized (if they are not already), and then those virtual machines (VMs) are spun up in the cloud as identical replicas of the on-premises data estate. This is also known as "lift and shift." Its main benefit is that it can be quickly completed, allowing the team to move from a "known" environment in the cloud to gradually incorporate new features of the cloud without having to maintain an on-premises data center. Rehosting is very useful for old applications that are still needed but are no longer being invested in. This strategy has the drawback that you still have all the drawbacks of an on-premises data center and few or none of the benefits of moving to the cloud.

5. **Refactor:** The refactor strategy focuses on gaining one or two of the most powerful benefits of cloud computing while also limiting the amount of new code needed for the migration. In a refactoring scenario, you'll typically see major architectural components of the on-premises application substituted for a better option in the cloud. For example, a refactored SQL Server + .NET application might migrate to Azure SQL Managed Instance to get rid of their

old custom-coded identity management and authentication module in favor of Azure's more robust Azure Active Directory identity management features. The team spends time writing more efficient SQL code as well as a new generation of the front end. Going forward, the enterprise then builds and deploys new features and apps using Azure DevOps Pipelines.

6. **Rearchitect:** The rearchitect strategy is similar to refactoring, except that it is more expansive in scope and usually alters one or more fundamental layers of the application stack. For example, you may choose to migrate the application stack to Linux-based containers (using Azure Container Instances) and managed by Azure Kubernetes Services (AKS) using the Azure Service Fabric and/or Azure App Service. With these all-new capabilities, your EA also specs out ways to amplify the value of the application by incorporating market sentiment analysis with an easy Twitter API feed processed by Azure Cosmos DB. Note: Rearchitecting and rebuilding strategies might also be called "modernization."

7. **Rebuild:** The rebuild strategy is the most arduous of strategies but is the most thorough way to migrate your database and application. When rebuilding, your plan is to throw away *most* although not all aspects of the old application. The motivation to rebuild is usually driven by major outside factors, such as a major new law that so completely alters the compliance requirements for your application that a major rewrite is in the works anyway. Other motivators for rebuilding include the elimination of a massive accumulation of technical debt, to dramatically reduce costs, or to facilitate vendor support on a critical business-critical application. When rebuilding, most EAs plan to reuse only the database schema and historic data stored in the old database.

8. **Replace:** This strategy might be called "Let's buy a SaaS product and turn off our in-house product." In this situation, the IT team cannot design and construct a better version than can be bought cheaply from an independent software vendor (ISV). In a situation like this, it's much easier to buy than to build; only now the IT team no longer has to worry about administration, maintenance, and hosting.

9.   **Implementation:** After completing the pilot project, the EA and team roll out the production application after making adjustments found during the earlier stages of the migration project. As we discuss later in this book, your EA (or you personally) will have many tools at your disposal to move data, databases, and other artifacts to the cloud. Which you choose depends on your requirements and which tool best satisfies those requirements. Implementation is typically either done as "turnkey," meaning the new production cloud environment is switched on as the old on-premises environment is switched off, or phased, meaning the new environment is gradually ramped up over time while the old environment is slowly wound down and then turned off.

10.   **Handoff:** Once production rollout is completed, the EA and team should conduct a postmortem on the project with an eye on future enhancements.

The EA should also have a vision for ancillary aspects of the migration, such as automation, orchestration, high availability, disaster recovery, service-level agreements (SLAs), and more.

Take this common database-centric example of backup and recovery. In many on-premises data centers, there are a lot of pockets of unregulated disk space. While usages of the SAN might be carefully managed, we still see DBAs to use the available storage of one of their local SQL Servers to speed up a recovery effort. Here's how this scenario often plays out:

1.   The DBA configures a nightly full database backup and 15-minute transaction log backups, which are written regularly to the SAN mount point provided to them.

2.   However, the SAN is busy and sometimes experiences bottlenecks on its Host Bus Adapters (HBAs). So the DBA also keeps seven days of those backups locally in case they need to recover a slightly older version of the database. The SLAs are tight, and every second counts.

3.   As the backups age on local storage, the oldest backup files are automatically deleted as the new files for the day are stored locally.

That works well on-premises because the DBA has full control over all resources and they have plenty of unused storage. In the cloud, the DBA would be headed for disaster. Why? Because you must pay for every bit of CPU and storage that you use. In this case, the DBA would basically have to pay for either 7 times as much storage, since that's how many daily copies of the backups they retain, or even 14 times as much storage, if they are using a high availability arrangement with multiple SQL Servers. OUCH!

In the cloud, the meter is always running.

The EA can help the DBA mix and match a variety of other Azure features to achieve a similar level of recoverability at a cheaper cost while also explaining the various trade-offs that are required in this scenario. For example, Azure offers many built-in replication and mirroring capabilities at the server and storage level, in addition to backup and recovery features incorporated into Azure SQL Database and Azure SQL Database Managed Instance.

Summary

Many migration projects are doomed from the start. And if not outright doomed, they are set up in advance for many trials and troubles. They now serve as cautionary tales. As we mentioned at the very beginning of this chapter, the lion's share of pitfalls relates to good leadership, good goal setting, good requirements, and expectation setting. Yes, technological issues can be quite difficult. But those roadblocks usually have ready answers and well-documented resolutions at hand.

But if you have not adequately prepared for the human element of a migration project, it's going to be a struggle. Some migrations that face too many obstacles and hurdles are faced to repatriate.

When your migration project moves from conception into the hands of operational IT leaders, make sure that their requirements are well defined. It is essential that they know what they're doing and that they are willing and able to work beyond their silos for the good of the whole team, the migration project, and the enterprise.

Otherwise, there could be in-fighting. There could be foot dragging. There could be interpersonal friction as middle managers seek to protect their turf and their hard-won technical relevance.

Another salient point to remember is that there is no replacement for employing a skilled enterprise architect or upskilling one of your current members of staff for the job. Beg, borrow, or steal those talented people, or hire them until you have built your own.

If you have to build your own, they'll have to go through a big process of planning that move, gathering all of the assessments that are necessary to make the migration project successful. They'll have to figure out all of the interconnections between all of the interconnected systems and the layers of dependency trees. Based on these factors, you may need to rehost, refactor, or even rebuild your application and databases over the course of your migration.

In essence, this book will train you for the role of a migration-centric enterprise architect. Are you prepared? Get ready to learn!

CHAPTER 2

Planning Considerations and Analysis

Migrating SQL Server workloads to Azure doesn't have to be risky, difficult, or expensive. With the correct approach, you can avoid many of the common pitfalls associated with the migration process, and those that can't be mitigated can be identified to allow for thorough assessment.

Although many project management frameworks can work well for migrating a database from on-premises to the Azure cloud, it's important to assess what works best for your organization. During that process of assessment, remember several important aspects that your project management framework needs to support. The framework you chose should address project scope, discover the data assets involved in migration, and execute testing in several important ways (e.g., building a performance benchmark of the on-premises database that will be migrated), validation, migration, postmigration operational assessment, and postmortem. All these stages require planning and preparation.

Topics Covered in This Chapter

We face two broad phases for assessment and planning. First, we must plan for moving the on-premises database or databases that we are moving to Azure. This phase is primarily focused on learning and documenting as much as possible about our on-premises databases. Second and equally important, we must plan for and implement the various Azure services and resources in order to properly host the migrated database. This phase primarily focuses on determining which Azure SQL platform is best for our migration and what, if any, additional Azure services we might need. Much like an aircraft flight plan, we might conceptualize these two major planning phases as "liftoff" and "landing".

© Kevin Kline, Denis McDowell, Dustin Dorsey, Matt Gordon 2022
K. Kline et al., *Pro Database Migration to Azure*, https://doi.org/10.1007/978-1-4842-8230-4_2

Scope Definition

It's possible to perform discovery throughout your entire IT estate to seek out all SQL Server systems when determining the scope of your migration. However, to successfully migrate workloads, it's best to restrict the initial project scope to that of a specific business system or department encapsulated within a single database or a few databases that interoperate together.

For example, you should be working with a clearly defined scope that fully defines the business requirements of any online transaction processing (OLTP), extract, transform, and load (ETL), and reporting layers. In addition, you should perform effective due diligence to ensure you haven't omitted any necessary systems or data and there aren't any departmental applications in use that you aren't aware of. It's both a discovery exercise and a validation of the initial scope.

The initial scope will feed into the basic migration plan and, subsequently, the discovery phase. As the migration progresses through the discovery phase, the scope can be refined based on the data collected. Likewise, as the testing phase occurs, the scope will likely be refined depending on whether it's viable to migrate workloads as is or if more work needs to be completed before a workload can be migrated to Azure.

You also want to consider how your business might evolve during the migration process. If standards or processes change, then it's important to understand if the migration project plan needs to reconcile with those changes too.

Tip   Refining your cloud migration scope requires flexibility and adaptability on your part. Your planning process should include multiple touchpoints to incorporate any changing conditions encountered during the migration process. However, it's equally important to avoid scope creep that adds extra work. It's better to start small and iterate through multiple versions, rather than delaying your migration project in the face of major scope creep. And be sure to disallow any requirement changes that cause scope creep without a compelling reason that's supported by empirical data.

Planning

When developing a plan for conducting a migration, you and your business leaders must understand that it's an iterative process. As the migration proceeds, new facts and data will become available that can impact the state of the plan. Addressing these changes is crucial. Trying to force the migration to adhere to the original plan without accounting for feedback will result in failure.

A combination of waterfall and Kanban project management methods works very well for managing migration activities, especially if your operations staff are managing the process. Should the migration be a larger, more formal project, then there could be a case for a combination of waterfall and scrum.

The waterfall method is used to control the phases rather than the underlying order in which the discrete activities take place. The overall process for a cloud migration would be logically broken down so that you complete the discovery phase before the analysis phase, for example. Within each of these phases, the use of the Kanban method allows your teams to tailor the approach to each system and complete the appropriate tasks.

Phase 1: Discovery

An in-depth understanding of your primary data systems helps you build a detailed migration plan. Key objectives for discovery include the following:

1. Systems from which the migration source consumes data

2. Systems for which the migration source produces data

3. Current versions of software components

4. Resource utilization (e.g., CPU, storage, memory) and performance benchmarks

5. Example workloads

6. Recovery Point Objectives (RPO) and Recovery Time Objectives (RTO)

7. Insight from application owners and business subject matter experts

You can collect some of this information using automated tools such as commercial monitoring or documentation software or free tools like the Microsoft Assessment and Planning (MAP) tool, available at `www.microsoft.com/en-us/download/details.aspx?id=7826`.

However, gathering other crucial information will require you to engage your business leaders, which has the added benefit of ensuring buy-in for the migration. Buy-in ensures that when you need to make requests from your business leaders, they're more inclined to collaborate with you.

Discovery of the Database

After you've completed a scan of your systems, you can then review the system configuration data about the servers and the databases they contain. Collecting this data allows you to build a data map or data catalog of your IT estate or at least the database(s) you plan to migrate. In addition, it's important to document data linkages within and across your database. For example, a SQL Server database might appear to be completely self-contained, but SQL Agent jobs directly related to your database are stored in the MSDB database, and logins for your database are stored in the Master database. Also, only after checking for linked servers does it become clear that it may be connected to many other databases.

Note   The Microsoft Assessment and Planning Toolkit, a.k.a. the MAP toolkit, can create system documentation and a data dictionary, greatly easing your ability to quickly document your IT estate. The Microsoft System Configuration Manager tool may also be useful, especially in enterprise IT environments. Other useful tools for this purpose include the PowerShell-driven utility, SQLPowerDoc, available from GitHub. You could also manually collect a variety of system details, for example, using the INFORMATION_SCHEMA system views. But this is a very cumbersome and non-automatable approach. ISV tools can be even better suited for these needs, but they are not free.

A data lineage map enables you to perform detailed impact analysis of any changes you'll make as part of the migration process. Visibility into the data flowing through each server, database, and SSIS package you'll be migrating allows you to understand the upstream and downstream impact of changes and the degree of entanglement your chosen database might face. You can also document and analyze the risks to the business for the migration activity.

In addition to collecting hard data related to servers, databases, packages, jobs, etc., you need to collect and catalog the metadata, which will be crucial to prioritizing activities, as well as ensuring the data and components are handled appropriately. By leveraging the data dictionary you have already captured, you can also tag sensitive data fields for PCI DSS, personally identifiable information (PII), General Data Protection Regulation (GDPR), HIPAA, or other compliance requirements.

You can also include extra metadata about application and database owners, points of contact, and what these entities are used for. The ability to add information about what an ETL job does, and more importantly why it does what it does, is key to understanding if the job is working as planned. Resource tags and classification can also be useful for future compliance efforts, especially once you have moved to the cloud. Azure Purview can make effective use of tagging and classification of your metadata.

A great use of the data dictionary is to add information around test criteria and what is a pass/fail. This information can then be used not only to build tests to validate the migration but also for postmigration business-as-usual (BAU) activities to support the continued development of your platform.

The work you do at this stage to capture the metadata for the databases to be migrated will live long past the migration process itself and provides your future self and colleagues with a treasure trove of useful information.

Discovery of the Business and Its Requirements

Another area of the discovery phase is understanding the business aspects of the migration process. You must consider what data is contained within the systems and databases, how important the data that is being migrated is to the business operation, and who is impacted by the migration. Be sure to ask for any documentation from earlier versions of the migration source. You might get lucky to find a serviceable document which only needs to be updated.

As part of the discovery process, you need to collect information about service-level agreements (SLAs) for performance, availability, and disaster recovery (DR) scenarios, including Recovery Point Objectives (RPO) and Recovery Time Objectives (RTO), as this information will drive the provisioning of the appropriate cloud storage tiers.

The information that you capture while speaking with your business leaders and users needs to be documented. This information can be added to your SQL Server systems via Extended Properties and annotations, in a wiki, or in a GitHub repository.

Discovery of the Workload

When performing an upgrade or migration, performance is a key metric the business will measure. Often, there's reluctance to adopt modern technology because of a perceived risk associated with being at the bleeding edge or even as a fast follower. However, you can establish the actual performance impact by providing greater visibility into performance and reliability by establishing a baseline.

In many cases, you can leverage modern technologies to improve performance or get the same level of performance with less resources. However, it's crucial to have a solid monitoring solution implemented to capture baseline data, which can be referenced during the testing and postmigration phases.

When baselining performance for a system prior to an upgrade or migration, it's important to identify the most important metrics to your business. Taking a blanket approach can result in too much data collected and important data points being lost in the noise. This is where benchmarking comes in, as it delivers built-in baselining capabilities and can collect all the key metrics for migration analysis.

When performing a cloud migration, the following elements can affect performance:

- **CPU utilization:** When analyzing CPU resources, it's important to take parallelism into account. Often, you'll hear about people trying to drive higher CPU usage by using fewer cores for a workload. Although this can be achieved, it's important to assess the impact on large queries that use multiple threads.

- **Disk I/Os:** Not all cloud storage is created equal. Although Azure premium storage might be solid state, it's typically network attached for the workload. Additionally, it's likely to be broken into performance tiers like a modern SAN. Furthermore, this component of your cloud infrastructure is further broken down into IOs/second and storage bandwidth. Many novices properly plan for storage requirements, since that is easy to measure across your enterprise servers and then provision in your Azure landing zone. However, one of the most common places we see migration projects fail is by not benchmarking I/O throughput on their existing systems and then provisioning inadequate disk I/O bandwidth in Azure. Thus, understanding your workload's use of storage, as both I/Os and storage, is vital when trying to achieve optimal or even adequate performance.

- **Memory utilization:** Memory in the cloud is tightly controlled. With IaaS solutions, you only get predefined configurations. With PaaS solutions, memory is abstracted away and not something you can affect directly. Knowing your memory utilization and associated relationship between data cache, such as page life expectancy (PLE), and storage throughput will help inform sizing decisions.

- **Query performance history:** In recent versions of SQL Server and in the Azure PaaS solutions, Microsoft has updated parts of the optimizer. This can result in behavioral changes both good and bad. Understanding query-level activity helps identify regressions that need to be fixed as well as performance improvements experienced from moving to a new system.

When capturing baselines, you must ensure you have performance data for all major time periods of your business, including key active hours for the business, special month- or year-end process (if possible), and any maintenance windows. (You must take maintenance periods into account because they're still required in the cloud. There's a misconception that these PaaS cloud solutions are entirely managed, but they aren't.)

Tip   Tools will help enormously. Pick the right tool for the job.

While specific assessment tools are covered in greater depth elsewhere in this book, a brief introduction is warranted here. Azure Migrate is really an "umbrella" of individual migration services in Azure. Those services are generally tailored to specific server roles (database server, web server, etc.) but can operate within the umbrella of Azure Migrate. Data Migration Assistant (DMA) helps assess a database's suitability for migration and ensure that no breaking changes or performance degradation will occur upon completion of the migration.

The SQL Server Migration Assistant (SSMA) provides tooling to migrate your Microsoft Access, IBM DB2, MySQL, Oracle, SAP HANA, and SAP ASE databases to SQL Server. Moving from one database platform to another is known as a *heterogeneous migration*. In these scenarios, SSMA greatly accelerates the heterogeneous migration effort. Note that SSMA is unlikely to successfully translate and migrate every single component of your source database and application. Table and schema objects definitely encounter few, if any, mistranslations between the source database to Azure SQL. However, it is fairly common to see about 20% of stored procedures and other programmable modules to require hand coding to successfully migrate. So although SSMA is very helpful, it can still mean that there is a lot of work to do even when using SSMA because it is not usually able to translate all elements of a database from one vendor platform to another.

Discovery Wrap-Up

Once you have collected the information, you have a sort of book of the database and application code base. It contains the information you need to move onto the testing and validation phase of the migration process. This topic is also discussed in greater detail in Chapter 6.

At a minimum for a migration source containing only a single application and a single database, you will want the message output of the *SP_HELP* system stored procedure for each schema object and programmable module. Programmable module is an SQL standard term for stored procedures, user-defined functions, triggers, views, and any other programmable such as a package on Oracle. You can gather this output by executing the T-SQL string *EXEC sp_msForeachtable "sp_help (?)"* within the source database to be migrated. You'll also want to use "EXEC *sp_help xyz*", where *xyz* is each programmable module in the source database. If you used any of the migration tools mentioned earlier, like DMS or SSMA, then you can simply use the reports they have created for you.

It also provides a means of conducting an effective database design review and data cleansing step of the process (see Chapter 6 for more details on data cleansing and validation). It's important to remember that this data needs to be kept up to date (another reason why using a discovery tool you can automate is ideal since you may have to resync due to new releases). A common reason that migration activities fail is because the project happens in isolation from the rest of the IT team and daily business activities, even while the source database and application are updated and changed. Unless the source code base is locked down and changes are prevented, you should frequently renew the database documentation to look for new schema changes and data feeds. Additionally, if new users are added to a system, the performance profile of the database systems might change, thus requiring new baselines.

All documentation created throughout the discovery phase should be considered living documents and maintained as such. Finding out about changes early is key to ensuring that they feed into the plan and result in a successful migration.

Phase 2: Landing Zone Testing and Validation

Once you clearly understand the database schema, data, and code within the scope of your migration, which we'll call the *source*, you need to know how it will behave on your new platform, which we'll call the *target* or *landing zone*. This is an ideal time to look at the source's transaction workload behavior on the target system to identify anomalies. This process of testing our landing zone, probably through multiple iterations, enables us to configure the ideal Azure production configuration which best balances performance, features, and cost. Although it is most important to identify and document performance degradation, it is also important to identify any improvements in performance you encounter. The reason is if your performance improvements are significant, you might have accidentally overprovisioned your landing zone, thus overpaying for the amount of resources your landing zone actually needs.

Note   With any luck, one or more members of your team are already acquainted with the Azure Well-Architected Framework. This is a set of guiding tenets that improve the quality of a workload in the Azure cloud. The five tenets of the framework are Cost Optimization, Operational Excellence, Performance Efficiency, Reliability, and Security. Incorporating these tenets into your migration process helps you create a high-quality, stable, and efficient cloud architecture. In a sense, this book is providing a wide set of experiences and recommendations that complement the framework. If you're unfamiliar with the framework, be sure to study it further at `https://docs.microsoft.com/en-us/azure/architecture/framework/`.

During this stage, you should evaluate migration options, start tinkering in Azure, and test the options that are likely to meet your needs. Although experienced staff will have preferences for how to move databases and other objects, it's important to keep an open mind about your migration options.

Finally, you'll need to define and set up the tests that properly validate both the condition of the migration source and the performance of the migration landing zone you have created in the Azure cloud. It's vital that decision points are defined and that the information needed to make decisions about proceeding or rolling back is available well ahead of the migration.

When possible, automation should be at the forefront of your testing processes. By automating tests, you can ensure that the output between test runs can be quickly compared and validated. Automated testing makes A/B testing of various landing zone options a fast and easy process. If you perform manual testing and validation, you risk that someone didn't follow the process correctly, which invalidates the results, leading to additional work and risk being added to the migration process.

Analysis and Testing

Now that you have documentation and performance baseline data in hand, you can refine the plan, build out the landing zone, and define the migration activities you need to undertake. You can break this phase of the migration into one or more work streams (see Figure 2-1), which are fed back into one another as you select the services and mechanisms you will use to migrate your databases. At each stage of analysis and testing, be sure to take detailed notes so that you have a comprehensive set of documentation showing your options, decisions, and outcomes during the migration process.

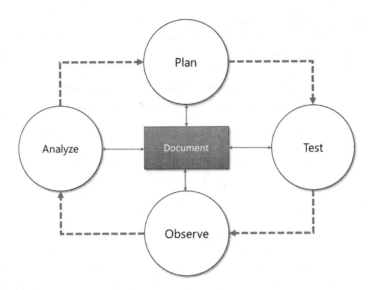

Figure 2-1. *Possible work streams for the analysis and testing phase*

Interview your business leaders to gain greater insight into their needs from the platform regarding service uptime, high availability (HA), disaster recovery (DR), and allowable downtime for the migration. Then, you can evaluate which Azure cloud services best align with those needs.

Availability and Uptime

Microsoft has uptime SLAs for Azure services, as shown in Table 2-1. However, in the case of IaaS VMs, there are certain configuration requirements you must adhere to for the SLAs to apply. These requirements will have an impact on the way that you design your cloud-based infrastructure. If your business needs higher uptime for your systems, then more complex configurations will be required to deliver it.

Table 2-1. *Uptime SLAs for various cloud services*

Azure Cloud Service	Requirement	Uptime SLA
IaaS VM	Two or more VMs deployed over two or more Azure Availability Zones within the same Azure region	99.99%*
IaaS VM	Two or more VMs deployed in the same Availability Set	99.95%*
IaaS VM	Single VM using premium managed storage for all Operating System and Data Disks	99.9%*
Azure SQL Database	Basic, Standard, and Premium tiers	99.99%**
Managed Instance	General purpose or business critical	99.99%***

*https://azure.microsoft.com/en-gb/support/legal/sla/virtual-machines/v1_8/
**https://azure.microsoft.com/en-gb/support/legal/sla/sql-database/v1_1/
***https://docs.microsoft.com/en-us/azure/sql-database/sql-database-managed-instance

For example, you might need to have two IaaS VM configurations spread over multiple Azure regions, in addition to deploying across Availability Zones within each region. Although each configuration still has the 99.99% uptime SLA, the chances of both configurations going offline at the same time are incredibly low. As with anything Internet related, there is always the risk of an outage. But it's important to understand the risk profile, costs, and benefits with each of your HA/DR options. Learn more about Azure PaaS HA/DR details at https://docs.microsoft.com/en-us/azure/azure-sql/database/high-availability-sla.

Phase 3: Choose the Migration Strategy

While there is plenty of nuance to choosing migration strategies for your applications, what is right for one app may not always be right for another. Consequently, we recommend choosing a strategy from one of two key paths.

Path one is commonly known as *lift and shift* or *rehosting*. Basically, we are going to move our servers, apps, and data to the cloud in largely the same form as they are on-premises. If each tier is on individual VMs, they will remain that way in the cloud.

They will remain constrained by the physical limits of the hardware they are on – though the cloud does offer some options for more easily increasing the hardware capacity of a VM without physically purchasing hardware and adding it to your current setup.

Note   This book emphasizes that your first strategic choice is whether to use a VM-centric approach (i.e., IaaS) or an Azure SQL-centric approach (i.e., PaaS). Once you've selected one or the other, you will then refine the selection. For example, you might choose Azure SQL Database instead of Azure SQL Managed Instance but then must decide which selection of Azure SQL service and tier is best for you.

Path two can be described as *modernization* or *rearchitecting*, although this term really doesn't do it justice or fully express the multitude of options for hosting data, apps, and their related services in the cloud. Perhaps you don't need a VM per physical database and can combine a number of them on a PaaS offering in the Azure SQL family. Do you use a lot of SQL Agent jobs or have no scheduled jobs? These factors all direct you to one choice of architecture or another. Depending on the architecture of your application and how to best leverage the features of your landing zone, there may be similar offerings for the application tier that removes your infrastructure team from managing server capacity, day-to-day firefighting, and installing servers. Instead, they can now focus on improving the performance of the application and cloud infrastructure itself.

Note   There is yet a third path, known as "recode," which is essentially a rewrite of an existing on-premises database and application. Bits and pieces of the original, on-premises code base might persist in the cloud version, but they are few and far between. This strategy is usually selected when an organization wants to reap all available benefits of the cloud and/or expel a massive buildup of technical debt within the on-premises code base. For example, your application might benefit from moving away from a monolith web service to a collection of microservices. However, since this path usually ends up reusing little or none of the original design, it is only a migration in the strictest sense of the word. Consequently, this book focuses on rehosting and rearchitecting strategies.

A simple and generic migration scenario is described in the images and text in Figure 2-2. These are certainly not the only possible scenarios, but it offers an idea of what the most direct migration option looks like. But if Figure 2-2 represents a migration scenario that could meet our requirements, that begs the question of whether it is our best option.

Should we simply lift and shift the application and underlying database? While Figure 2-2 represents an example on-premises scenario that uses VMWare and would leverage a PaaS Azure SQL offering with Managed Instance, we are still relying on VMs for the web and application tiers. Azure SQL Database would cost less and offer greater flexibility as the workloads change, but the database is not ready for the fully managed nature of Azure SQL Database. That means the team must choose between a VM-centric migration strategy (IaaS) and Azure SQL Managed Instance (PaaS). Is this truly modernization or perhaps just a baby step toward it?

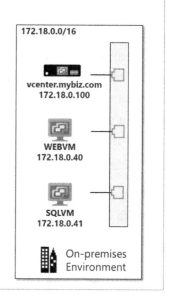

Rehost a.k.a. "Lift and Shift" to Azure IaaS

Traditional three-tier application running on-premises on Windows & SQL 2008R2 in VMware

Business Goals of Azure Migration:

Same performance capabilities as experienced on-premises in VMWare

The team doesn't want to invest in the app, Mybiz will simply move the app safely to the cloud

The database doesn't meet the requirements of Azure SQL Database, even though they want to use a PaaS solution. Will have to use IaaS.

172.18.0.0/16

vcenter.mybiz.com
172.18.0.100

WEBVM
172.18.0.40

SQLVM
172.18.0.41

On-premises Environment

Figure 2-2. *Lift-and-shift example scenario*

While rehosting may seem right for some of your apps and environments, it is arguably not the right decision for most of them given the cloud options available, particularly within the data platform. As shown in Figure 2-3, there are additional options beyond simply running SQL Server within a VM. Depending on the business requirements of your migration, you may wish to use Azure SQL Database Elastic Pool or Azure SQL Database Managed Instance.

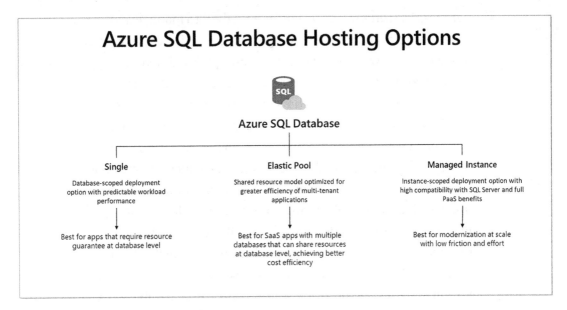

Figure 2-3. *Azure SQL hosting options*

While Figure 2-3 shows the primary Azure SQL purchasing options, it is worth paying particular attention to some other of the specialized offerings in the realm of Azure SQL:

- *Azure SQL Database (single)* is a fully managed Platform as a Service (PaaS) database that negates the need for you to manually administrate the service. You won't have to worry about backups, patches, upgrades, and monitoring. Your Azure SQL Database applications always run on the latest stable release of the SQL Server engine with a 99.99% availability service level. Azure SQL supports features of an on-premises SQL Server at a database level but does not support server-level features like SQL Agent, Linked Servers, SQL Server Auditing, and DB Mail. Generally speaking, you pay for Azure SQL using a *vCore-based billing model* or a *DTU-based billing model*. The vCore billing model lets you choose the number of CPU cores, memory, and the speed and amount of storage. Alternatively, the DTU (Database Transaction Unit) purchasing model blends compute, memory, and I/O resources into three service tiers that support light, medium, and heavy database workloads.

- *Azure SQL Database Elastic Pool* is a method to manage and scale multiple databases in a simple and cost-effective way. You might call this Azure SQL Database (multiple) to differentiate it from standard Azure SQL. For example, an ideal scenario for Elastic Pool is an SaaS ISV who also wishes to act as the hosting provider for their many customers, each with their own database. But some customers have a need for high performance, while others need only a limited amount of performance. Elastic pools satisfy this requirement by ensuring these many databases get the resources they need within a predictable budget. Users of Elastic Pools are not billed on a per-database charge but are instead billed for each hour the pool exists at the highest rate of usage by eDTU or vCores, even if you use a fraction of an hour. Within your defined pool of resources, individual databases may expand or shrink the amount of resources they consume, within set parameters.

- *Azure SQL Database Hyperscale* service tier provides a high-performance computing model to the Azure SQL space. While other Azure SQL service tiers of *General Purpose* (a.k.a. *Standard*) and *Business Critical* (a.k.a. *Premium*) have a cap on database size of 4 TB per database, thus making it unsuitable for customers who are migrating to a very large database. Hyperscale supports up to 100 TB. That in and of itself may not seem very exciting, but Hyperscale also represents a completely redesigned storage architecture for Azure SQL. Without delving deeply into the technical details, it leverages different tiers of Azure storage to ensure that log throughput is extremely robust. Beyond that, and most excitingly, because it uses Blob storage for backups and restores, working with the backups and restores for very large databases goes from hours to seconds in nearly every workload scenario.

- *Azure SQL Database Edge* represents the opposite end of the size continuum from Hyperscale. Edge is intended for IoT usage, so it runs on a very small footprint – meaning you do not have to consume a lot of resources and incur the overhead of a VM for very small databases. Edge also allows time-scale analysis and supports a substantial portion of the T-SQL surface area of more conventional deployments of SQL Server. Although Edge is unlikely to appear in a migration scenario, it is a possibility. Your technical team can leverage this in scenarios where an IoT-scale database is needed to run on hundreds or even thousands of nodes, such as the RFID tags on a fleet of delivery vehicles.

- *Azure SQL Database Serverless* is a single database compute tier that automatically scales compute-based workloads on demand and bills for the amount of compute used per second. In a nutshell, databases running on Serverless are intentionally configured to reduce your costs. It can even be configured to automatically shut down the service when no one is using the service, thus offering significant savings for databases that have working hours and off hours, like a traditional retail Point-of-Sale terminal. While Serverless databases are offline, you are billed only for the storage costs of your database.

- *Azure SQL Managed Instance (MI)* is your landing zone choice when you want a feature set that is as close to on-premises SQL Server Enterprise Edition as possible. For example, MI supports cross-database transactions and Service Broker, while Azure SQL does not. The flipside is that you must also do more administrative work, such as index tuning or taking and restoring database backups. Like Azure SQL, MI includes a 99.99% service-level agreement as well as running on top of the latest stable release of the SQL Server engine. MI further includes built-in high-availability, compatibility-level protections.

All of that said, if you are looking for one single recommendation from this chapter for a "one-size-fits-all" migration destination for all your SQL Server databases, there isn't one. But if there is one Azure SQL offering that comes nearest to meeting the needs of most migration projects, it is Azure SQL Managed Instance.

It very closely resembles Enterprise Edition of on-premises SQL Server when you connect to it (giving your DBAs a level of comfort) and it offers the benefits of a PaaS service, since it is one, giving your DBAs and operations team a level of comfort that it will be well maintained with proper care and feeding. The only significant technological blocker that we have encountered is that it does not support filestream and filetable features from on-premises SQL Server. Its one noticeable downside is cost. But if you take advantage of the advice throughout this book, then the initial price disadvantage of MI quickly becomes very competitive with what you spend for on-premises SQL Server.

Identifying and Configuring the Migration Landing Zone

We use more than one term meaning the destination of your Azure migration, such as migration target, target environment, and landing zone. Going forward, we may use any one of those, but in all cases, we mean the set of defined services, service tiers, and functionality of the Azure cloud we use going forward. As we discussed earlier, there are many different options when it comes to moving your databases to the cloud. Each option has slight nuances to how it should be used as well as outright blockers to adoption.

Working through these options manually can be difficult. However, Microsoft offers tools that can help you automate the selection process and size the target platform. Let's talk about those tools in the next section:

Data Migration Assistant

The Database Migration Assistant (DMA) is a free tool from Microsoft that performs analysis of our source system to identify potential blockers when migrating an on-premises SQL Server. It can be used for migrating databases from one SQL Server system to another, to Azure SQL Database, or to Azure SQL Database Managed Instance.

The DMA performs analysis on both the instance and database level of your source system to identify issues that need to be resolved before you migrate. Using the DMA on the systems that you have defined as within scope for the migration and performing an analysis yield results like those shown in Figure 2-4.

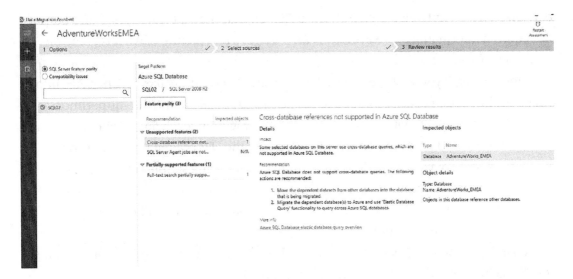

Figure 2-4. *An example DMA assessment identifying incompatibilities between an on-premises SQL Server 2008 R2 source and an Azure SQL Database landing zone*

You can see in Figure 2-4 that the DMA points out issues with feature parity and compatibility issues, as well as features currently in use on the source database that are entirely unsupported by Azure SQL or only partially supported. As you click through the report, you are also provided with details about how to remediate or work around the issues found by the DMA. It's not uncommon in your migration process to identify inconsistencies and blocking conditions, make changes to fix those issues, and then to rerun DMA for a new assessment. Lather, rinse, repeat until your source database is finally in migration-ready condition.

Sizing the Cloud Systems

When sizing the cloud systems, there are several options, including free tools from Microsoft, available to derive a starting point.

Azure SQL Database DTU Calculator

The DTU calculator, a free tool built by engineers at Microsoft, takes a CSV of performance counters collected using Windows Performance Monitor (PerfMon) and then calculates how many DTUs your Azure SQL Database would need. There are links to a PowerShell script or a command-line utility on that site that can capture and output the required information. However, you can only get a real picture of what is going on with a longer-range capture.

Interestingly, if you visit the DTU calculator site and want to take a deeper dive into what PerfMon counters are compiled to feed to the DTU calculator, the sql-perfmon. ps1 script referred to at that site is well worth spending time reviewing. While modifying it is not recommended (as then it will not work with the DTU calculator), it may plant the seed for some of your own ideas for your own monitoring scripts you can use to supplement whatever other scripts and tools you have in your environment. Reverse-engineering something almost always teaches you something you did not know previously, and this is yet another case of that.

To gain a deeper understanding of the DTU performance metric, check out the blog post "What the heck is a DTU?" by Andy Mallon (t|b).

One final note, as of publishing time, Basic, Standard, and Premium service tiers are available within the DTU-based purchasing model. General Purpose, Hyperscale, and Business Critical service tiers are available under the vCore-based purchasing model. Read more about storage differences under the DTU purchasing model at `https://docs.microsoft.com/en-us/azure/azure-sql/database/service-tiers-dtu#compare-service-tiers`. For example, there is a spinning disk tier and a higher speed SSD tier for DTU-centric purchasing.

vCores Licensing Model

The vCore purchasing model is a more recent addition as a second purchasing model for Azure SQL services. It offers multiple benefits. First, you can bring your own licenses, if you have already purchased them from Microsoft. You can also use the licenses for on-premises SQL Server to exercise the *Azure Hybrid Benefit* licensing option (refer to Chapter 3 for more details). It can offer more flexibility to you as you choose your Azure SQL–based migration targets. But with that flexibility come some extra complications as well. Compare vCore licensing to the DTU model where you purchase a bundle of compute and storage resources underneath the Azure SQL label. With the vCore purchasing model, you purchase compute resources independent of storage resources (which can be an important thing to point out to the people that pay the bill). With that unbundling of compute and storage comes the additional option of choosing between storage tiers and compute tiers.

In a sense, you can mix and match to achieve your best balance of performance versus cost. On the compute tier, you may choose between a provisioned and a serverless compute tier.

A provisioned compute tier means that the compute resources you have provisioned will remain yours until you delete the Azure SQL resource. Serverless simply means you define the minimum and maximum compute resources needed for your workload and Azure will, within configurable guidelines set by you, autoscale those resources to help you manage both cost and performance. It also allows you to pause (and resume) the database based on those autoscaling and workload guidelines you've established.

Database Migration Assistant

The DMA not only performs analysis on migration blockers but also includes a PowerShell script that can be used to help size target systems based on performance analysis data. The script takes several inputs to help provide an estimate on sizing.

DMA can capture the data it needs to perform this operation. However, this capture is typically a small set of data that isn't always representative of the true workload. By using a monitoring solution to track performance over a longer period, it's possible to leverage this tool to provide the analysis engine with more data. This, in turn, provides a higher degree of trust and accuracy in the sizing data that will be returned by DMA.

IaaS VMs

Sizing Azure IaaS VMs is more akin to the traditional way that you size on-premises systems. However, there are a couple of concepts that you need to understand about storage and how you approach cloud systems to ensure that you select the right size to host your workloads.

Azure Storage

There are multiple types of Azure storage available, from Standard to Premium disks, both managed and unmanaged. The type of storage that you select for your VMs will have a bearing not only on performance but also on whether you can meet the performance requirements of your users and perhaps better maintain adherence to an organizational SLA.

Should you simply create your on-premises storage patterns in the Azure cloud? Probably not. After all, if you completely reproduce your on-premises configuration in the cloud, you can only get from it what you have already gotten on-premises. In other words, what you got is what you will get – without any of the potential benefits and optimizations offered by Azure.

Here are some tips when considering this question. For files that live in the SQL VM, you only ever want to use managed disks unless you have the lightest of transactional workloads. This choice also ties into the availability model you'll use in the cloud. For database workloads, you are likely to only ever run on Premium storage or even Ultra Disk.

The size and class of VM selected will also dictate the number of data disks that can be used and the performance profile for them regarding IOPS and throughput. Microsoft's recommendations for running SQL Server workloads on Azure VMs are available in the "Performance guidelines for SQL Server in Azure Virtual Machines" documentation article.

Adopting a Cloud Mindset

When you provision on-premises systems, you do so by estimating the life span and factoring in growth over time, which often means you will overprovision capability for the system just in case. If you take this approach with cloud systems, it will cost more money than you need to spend, ultimately reducing ROI.

When working with cloud-based systems, whether IaaS or PaaS, you want to run with an overall higher resource utilization. Whereas you would see many on-premises systems running with 30% to 40% utilization for CPU, in the cloud, you want to push CPU utilization as high as you can go while allowing for performance spikes. Doing so minimizes additional spend on resources that aren't being used. It's easy to plan for scaling up a VM from one class to another, and it's just as easy with PaaS to move from one service tier to the next. Although these operations are typically offline events, with effective planning, you can minimize the impact to end users.

Analyzing Workloads

A crucial part of the testing and validation phase is ensuring that when databases are moved to cloud-based systems, they perform as expected. This analysis isn't limited to hard performance counter metrics such as transactions per second, I/O throughput, and response times. It also includes confirming that queries return the correct results and that the correct data is being written to your databases.

Workload analysis should be completed in several iterations, starting with the most basic tests, to ensure that applications can connect and simple transactions can be processed. In many cases, you might need to perform an action on the dev/test version of the system on-premises, repeat that action on your new platform, and then compare the results.

Validating data can be a painstaking task – automation should be applied whenever possible. Writing formal tests in code at this stage will greatly help when performing more in-depth workload testing. Also, automating the data validation process ensures that tests are repeatable, you can trust the output, and test runs can be compared to one another. When you manually run tests at scale, small discrepancies can creep in and invalidate the results.

Workload Replay Options

When performing workload testing, it's best to make use of a production workload, which allows you to validate the new systems against a known system and production baselines to ensure parity. Capturing and replaying workloads isn't a simple task. However, there are several options available to help with this task. These include OStress in the RML Utilities suite from Microsoft or SQLWorkload in the WorkloadTools open-source project by Gianluca Sartori (t|b). These tools have similar functionality but complement one another.

WorkloadTools

WorkloadTools is useful for live streaming a workload from one server to another. If you leverage Extended Events (XE), workloads can be streamed from one server to another with low overhead. It's also possible to perform workload streaming via SQL Trace on earlier versions of SQL Server where there isn't parity between SQL Trace and XE for events.

You can also capture a SQL Server workload with SQL Trace and output to Trace files. These files can then be converted to a replay format that SQLWorkload can process. The advantage of this approach is that it lets you capture several different workloads and store them for repeated use throughout the analysis phase as well as a baseline.

OStress

OStress is a replay tool that has been around for a long time and works on both SQL Trace and XE files. The process for capturing and converting a workload with OStress is a bit more complex than SQLWorkload. However, OStress includes an additional capability for the replay mode: multiuser simulation for testing concurrency limits. You can achieve this by combining OStress with ORCA in the RML Utilities suite.

The workload is captured using native XE or SQL Trace to files; these files are processed into the Replay Markup Language (RML) using the ReadTrace tool. OStress then reads the RML files to replay the workload against a target database.

OStress can be used in conjunction with the ORCA tool in RML Utilities, making it possible to coordinate the replay of the workload over multiple clients. This approach allows you to simulate a multiuser environment to determine if there are potential concurrency issues when the workload is run on the new system.

Workload Replay Process

The most crucial element of the workload replay task is ensuring a consistent database that can be reset on the target systems. You should take a full backup of the source database(s) for the workload that you will capture. Then, create a marked transaction right before starting the workload capture, which allows you to take a transaction log backup and restore the database to the marked transaction, after which you can replay the workload.

Note   Microsoft's Database Experimentation Assistant (DEA) can help with workload replay. However, there's licensing overhead for this tool due to the reliance on the SQL Server Distributed Replay utility.

This assumes, however, that you're working with a full or bulk-logged recovery model. If you're using simple recovery, it can be more difficult to coordinate the backup with workload capture. In this instance, a potential solution would be to trace the backup as well as the workload and then manually modify the workload records to remove the ones prior to the completion of the backup. However, this isn't an exact science and can be more difficult to complete.

When replaying the workload, the database should be restored, workload replayed, and system monitored, as shown in Figure 2-5. This monitoring session can then be compared to the baseline established during the workload capture or discovery process baseline. This database can be restored, and the workload replayed as needed, based on different configurations for the target system.

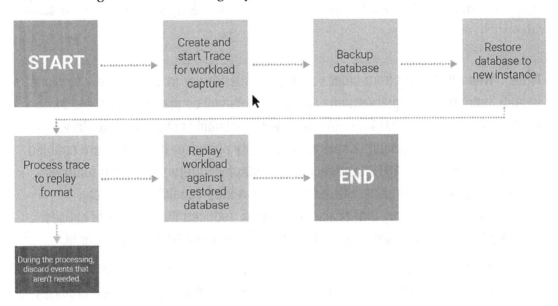

Figure 2-5. An example workload replay process

For more information about how SQL Server backups work so that you can decide which events might need to be discarded from the trace based on the transactions in the backup, see Paul Randal's blog posts "More on how much transaction log a full backup includes" and "Debunking a couple of myths around full database backups."

Monitoring for Workload Analysis

When performing workload analysis, you're likely to move through several iterations as you tune the configuration and make changes to code to work around any issues or incompatibilities. Ideally, you have a high-quality ISV monitoring system made specifically for SQL Server database performance monitoring. However, you can make due with Windows Performance Monitor (PerfMon) to collect a historic log of the most important performance counters over time. Monitoring the source database allows you to create baselines for specific periods of interest. The detailed performance metrics that are captured also enable you to track resource utilization over time for many counters. This collection of performance telemetry enables you to identify problematic large queries and small but frequently executed queries, which, in aggregate, cause problems with overall resource usage.

Note   When migrating from older versions of SQL Server to PaaS or new versions in IaaS, note that new query engine improvements will be present. Here are some examples. SQL Server 2014 introduced a new cardinality estimator. SQL Server 2016 altered several default behaviors, and query optimizer hot fixes were enabled by default. SQL Server 2017 introduced new adaptive query processing features to optimize query recompiles. SQL Server 2019 added capabilities that can dramatically alter the way that scalar UDFs work, reduced the number of causes for stored procedure recompiles, and offered improved query memory grants.

All these changes will be available in PaaS offerings ahead of on-premises SQL Server versions. Take the time to understand the potential effect of these changes to how your queries and workloads function as part of any migration effort.

With your performance telemetry from the source database, you can now monitor the new target landing zone to compare against your on-premises baselines. The deep insight into the way the workload is executing and the execution plans in use allow you to identify both areas of improvement due to the latest version of SQL Server in play. Or, more importantly, performance regressions that need to be resolved to ensure end users aren't affected and system performance isn't compromised.

Validating the Data

In addition to validating the performance and behavior of workloads on the SQL Server source and target, you must ensure that data is written to the database correctly and query output is as expected. This can be a complex and repetitive task that is best suited to automation rather than human operators. (We cover this in greater detail as an important independent step of the migration process in Chapter 13.)

By clearly defining the data validation tests for a predefined source database, you can ensure that with every workload run, you will validate the data written to the database and spot any data quality issues. This is important for two reasons:

1. The rate at which bugs are fixed in SQL Server has accelerated dramatically. This means that any behavioral changes introduced with these fixes could manifest in the results you see returned by your queries. Automated testing to ensure that data is consistent on both source and target helps identify these changes before they become a problem.

2. Microsoft has altered the default behavior of the SQL Server engine. For example, in SQL Server 2016, there were changes to the level of accuracy for values returned when there were implicit conversions between numeric and datetime types. Depending on the tolerances in your system, these variances could be something that you need to detect and then update code to maintain behavior.

Microsoft is very diligent when it comes to documenting changes between versions and editions and making that information available. Changes between versions of the database and of the code base are a typical root cause for data validation issues. Since it's possible to see the breaking changes to SQL Server database engine features in the Microsoft documentation, you may be able to avoid an extensive testing cycle if Microsoft has made no changes for a while. On the other hand, we strongly recommend that you plan for this type of data quality and validation testing.

Platform Capability Validation

Although you must understand the performance characteristics of the workload and ensure data validation, the operational manageability of the cloud platform must meet the business needs that you documented during the discovery phase.

Migrating from an on-premises system, where you have a high degree of control over how backups and maintenance operations are undertaken, to a cloud platform, where many of these actions are abstracted or handled by the platform and are therefore outside of your control, requires that you adapt your operational practices.

High Availability Configurations

Azure PaaS solutions such as Azure SQL Database and Azure SQL Database Managed Instance have HA features built in. Therefore, you must cede control to the platform and trust that failovers will be as advertised because there's no way to trigger and validate a failover event to understand the effect on your applications. However, IaaS solutions rely on you to set up the appropriate configuration.

The natural choice for HA in Azure VMs is AlwaysOn Availability Groups, especially given their support for DTC transactions and functionality historically available only in Failover Cluster Instances (FCI).

The implementation of Availability Groups in Azure is largely the same as an on-premises implementation, but there are two crucial differences. First, the management of resources, such as IP addresses, and network firewalls are controlled outside of the OS. Second, network traffic redirection for the listener requires the use of Azure Load Balancer services. (Note: This isn't required for newer CUs of SQL Server 2019 running on WS 2019+ as described at `https://docs.microsoft.com/en-us/azure/azure-sql/virtual-machines/windows/availability-group-load-balancer-portal-configure?view=azuresql`.)

These differences mean that operations staff must have a deep understanding of the Azure cloud platform as well as of SQL Server; otherwise, the operational capabilities of the solution can be compromised.

Disaster Recovery Scenarios

When it comes to DR scenarios, it doesn't matter whether it's a PaaS or IaaS solution that you're deploying – there's still an onus on operational staff to understand how to perform a recovery operation. There are two key elements to DR planning when it comes to cloud systems. These elements, described in greater detail in the following, can be used individually or in combination.

Backup Retention and Recovery Scenarios

Restoring these managed backups isn't the same as performing a native restore. Therefore, it's strongly advised that you test and retest the restore functionality during this phase of the testing process. Where possible, multiple full DR tests should be undertaken, not only to familiarize the operations staff with the new way of working but also to highlight any issues with the platform implementation that need to be resolved ahead of the final migration.

When using Azure SQL Database and Managed Instance, the backup process for your databases is handled by the Azure platform. By default, Azure SQL Database and Managed Instance have transaction log backups taken on a schedule between every five and ten minutes, respectively. Depending on the service tier selected, backups will be retained for one to five weeks. This schedule allows for point-in-time recovery for databases deployed to these services within a specific region. If you require longer-term backup availability, then you will need to configure Long-Term Retention, which allows for backups to be held for up to ten years.

The diagram in Figure 2-6 shows the Azure SQL Database restore process for a single database deployed to a server in an Azure region.

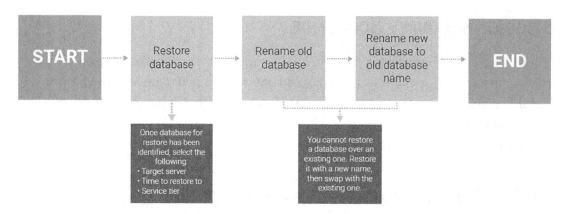

Figure 2-6. *Process for restoring an Azure SQL Database*

The step-by-step process shown in Figure 2-6 is a logical and intuitive one. Even better than manually performing this workflow is to automate it. We're in luck. Restore operations for Azure SQL Database can be easily scripted with PowerShell, allowing for restore operations to be automated as needed and for effective testing, as shown in the following example code:

```
## Declare Parameters
$SourceDatabaseName = "Accounts"
$ServerName = "s1dbwp01"
$SourceResourceGroupName = "SentryOne-WP-RG"
$RestoreTargetDatabaseName = "Accounts_Restored"
$PointInTimeToRestore = (Get-Date -Year 2019 -Month 05 -Day 28 -Hour
18 -Minute 17 -Second 00)

## Get Azure SQL Server
$SqlServerParams = @{
    Name = $ServerName
    ResourceGroupName = $SourceResourceGroupName
}
$AzureSqlServer = Get-AzSqlServer @SqlServerParams

## Get Azure SQL DB we want to restore a version of
$AzureSqlDbParams = @{
    Name = $SourceDatabaseName
    Server = $AzureSqlServer.ServerName
```

```
    ResourceGroupName = $AzureSqlServer.ResourceGroupName
}
$SourceDatabase = Get-AzSqlDatbase @AzureSqlDbParams

## Restore to new database on same server
$RestoreParams = @{
    FromPointInTimeBackup = $true
    PointInTime = $PointInTimeToRestore
    ServerName = $AzureSqlServer.ServerName
    ResourceGroupName = $AzureSqlServer.ResourceGroupName
    TargetDatabaseName = $RestoreTargetDatabaseName
    ResourceId = $SourceDatabase.ResourceId
    Edition = $SourceDatabase.Edition
    ServiceObjectiveName = $SourceDatabase.CurrentServiceObjectiveName
}
Restore-AzSqlDatabase @RestoreParams

$OldDbRenameParams = @{
    DatabaseName = $SourceDatabase.DatabaseName
    ServerName = $SourceDatabase.ServerName
    ResourceGroupName = $SourceDatabase.ResourceGroupName
    NewName = ($SourceDatabase.DatabaseName + "_old")
}
Set-AzSqlDatabase @OldDbRenameParams

## Rename new database to have old database name
$NewDbRenameParams = @{
    DatabaseName = $RestoreTargetDatabaseName
    ServerName = $SourceDatabase.ServerName
    ResourceGroupName = $SourceDatabase.ResourceGroupName
    NewName = $SourceDatabase.DatabaseName
}
Set-AzSqlDatabase @NewDbRenameParams
```

When it comes to DR planning for IaaS VMs, the process is remarkably similar to on-premises deployments. Your operations team would schedule native database backups to your specific Azure storage and configure the appropriate clean-up routines.

> **Tip**   Many organizations are ignorant that a common best practice is to make use of the native backup to URL functionality within your SQL Server VM. In this scenario, backup to URL not only eases management issues but also enhances resiliency by allowing you to replicate your backups to another Azure region. That way, you can avoid regional emergency situations, like floods, tornados, wildfires, and the like.

In addition, you may optionally choose to spend a bit more money so you can use the Azure Backup Service for SQL VMs, an enterprise backup feature set that enables you to centrally control large numbers of SQL VM backups, point-in-time recovery, and customizable retention policies. We recommend you familiarize yourself with all of the choices for Azure SQL VM backup and restore at `https://docs.microsoft.com/en-us/ azure/azure-sql/virtual-machines/windows/backup-restore`.

Geo-replication and Multiregion Deployments

One facet of cloud usage that many organizations neglect to account for is cloud region failures. All major cloud vendors have had outages that have impacted entire regions. Many of the Azure services have geo-redundancy capabilities that you can enable. Azure SQL Database and Managed Instance can leverage auto-failover groups for multiregion failover capability.

Auto-failover groups appear very similar to the Availability Group Listener for o n-premises SQL Server, providing an abstraction layer for multiple replicas. Planning for a multiregion solution can help improve the resilience of the platforms you build in the cloud. To learn more about auto-failover groups, see the Microsoft article "Use auto-failover groups to enable transparent and coordinated failover of multiple databases."

When building IaaS solutions with Azure VMs, it's highly advisable to use existing patterns and practices for multisite deployments. Again, you can use Availability Groups and Distributed Availability Groups within SQL Server. More information on these configurations for High Availability and Disaster Recovery on Azure VMs can be found in Microsoft's "High availability and disaster recovery for SQL Server in Azure Virtual Machines" documentation article.

Migration Dry Run

Toward the end of the analysis phase, you should be ready to test the methods that you will use to migrate your workload from on-premises to the cloud.

As with anything, there are multiple ways to achieve the objective. However, rather than immediately falling back to tried-and-tested methods that you're familiar with, you need to evaluate all the options and make a final decision based on gathered data and facts to ensure the migration methodology is appropriate and meets the needs of the business.

When migrating SQL Server workloads from on-premises to Azure, there are several options, as illustrated in Figure 2-7.

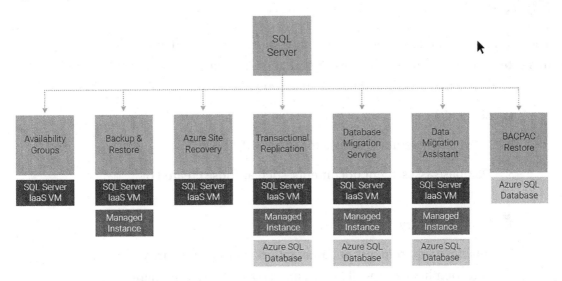

Figure 2-7. *Alternatives when migrating on-premises SQL Server to Azure*

As shown in Figure 2-7, there are many ways to move your database or data to Azure. And the choices shown in the figure do not even include the easy-to-use wizards available in SSMS and ADS.

The preferred method of moving an on-premises SQL Server database to Azure is to simply take a database backup and then restore it to the Azure landing zone. This approach is sometimes called the *turnkey* method. However, this approach only works for two target platforms: Azure SQL Managed Instance and SQL Server in Azure VMs. If you are migrating to Azure SQL Database, you will have to use another method. Instead of backup and then restore, you might wish to use a BACPAC and then apply that to an Azure SQL Database.

Another situation where backup and then restore cannot be used is when your requirements dictate that the source and target systems must run in parallel during the time needed to conduct a phased migration. In this scenario, we recommend that you use the Azure Database Migration Service (DMS). You could also use a data movement solution, such as transactional replication.

Finally, if you have a very large VM environment, you have another option to lift and shift all of your on-premises VMs at once using Azure Site Recovery. This is not only a great way to improve the recoverability of an existing Azure IaaS estate, it's also a great way to lift and shift all at once.

Summary

The analysis and validation phase of the migration effort is vital. If you don't complete this activity, there's a higher degree of risk that the migration effort won't succeed.

The outputs that you should have in place are as follows:

1. Scope definition

2. Migration inventory, using a tool like MAP Toolkit

3. Database assessment documents, such as those created by DMA or DMS

4. Data dictionary

5. Business requirements documentation, including security and compliance, as well as backup, recoverability, and high-availability requirements such as those described in RPO and RTO agreements

6. Performance benchmark of on-premises SQL Server

7. Analysis from benchmarks that identify the landing zone requirements, for example, SQL Server in Azure VMs server-level and performance tiers or number of vCores, memory, and storage for Azure SQL Managed Instances

8. Remediation plan for any findings from DMA and/or DMS

9. Test automation for validating the success of the migration checkpoints to provide details to the decision makers for continue/rollback quality gates or user acceptance testing and assessments of performance in the landing zone

10. Actual migration

11. Platform monitoring and analysis in place for source and destination system to ensure user expectations meet the needs established in the performance benchmarks

12. Final sign-off and switchover

While each of these steps is described in some detail in this chapter, the remaining chapters in this book go into greater detail on one or more steps in the preceding list. For example, Chapter 10 of this book is all about step 10 in the preceding list. Steps 2–4 in the preceding list correlate to Chapter 9 of this book. Read on for deep dives into these topics and many more.

CHAPTER 3

Budgeting for an Azure Migration

When companies go through evaluations for an Azure cloud migration, one of the most important considerations that come up is being able to understand and manage the cost. While there are certainly some companies willing to migrate despite the costs, this is not going to be the case with most. These companies need to get comfortable with the costs associated and have an understanding of what to expect before making any commitment.

Throughout our careers, we have worked with several companies teaching about the benefits of the cloud and explaining why it is the right solution to meet their needs. However, a common theme we find is that cost was almost always one of the primary limiting factors. In most situations, it was not even because it was so expensive; it was just because they did not know what to expect. So even though they would agree it was the right solution, they stuck with their existing on-premises setup since expectations could more easily be set and adhered to.

And it makes sense. Do you go to buy a new car and find the one you love and just say you want it without ever looking at the price tag, quoting insurance, or discussing terms of the loan? If you have enough money, you may not, but for most people, this is absolutely something they want to understand.

Cost management is often broken into two components. The first is *how to budget for an Azure migration*, which will be the focal point of this chapter. Here, we will focus on what you need to know and do to get started answering the question of what you should expect for cloud costs. We will then look at how you take this information and apply it to create a budget for your migration. In the next chapter, we focus on the second part, which is *managing your cloud costs* once you are in the cloud. In that chapter, we will focus on how you control your cloud costs post migration and adhere to the budgets you create.

© Kevin Kline, Denis McDowell, Dustin Dorsey, Matt Gordon 2022
K. Kline et al., *Pro Database Migration to Azure*, https://doi.org/10.1007/978-1-4842-8230-4_3

Why Should You Care?

Before we start to dive into the details of budgeting, it is worthwhile to spend a little bit of time talking about why you should care about this subject. For many of the technical readers of this book, you may see this chapter and overlook this content thinking it is not applicable or your responsibility. Well, you would be wrong. In the past, this may have been true, but the cloud has changed the way technical professionals must think, and this is becoming an important skill set for the modern cloud engineers and architects. Your current title may not reflect one of these roles, but guess who the people are that are getting them? They are people like you who come from infrastructure, networking, and database administrator backgrounds.

Most successful companies put a lot of emphasis on keeping their costs down as much as possible. Managing and understanding costs is important for them; thus, it should be important to the people that work for them. When looking at a big project like a cloud migration, cost is statistically one of the leading factors that block progress and can halt the entire project.

But what you may not be as familiar with is the number of people who are migrating away from the cloud due to the costs ending up being higher than expected. This is often referred to as *cloud repatriation,* which is a term that describes migrating from cloud back to on-premises and is a growing trend among organizations. There are several factors that go into making this decision, and one of the top reasons is due to cost. Incidentally, one of the key reasons driving repatriation is that those conducting the migration have failed to plan appropriately.

Companies now realize that there are cases where they can run workloads cheaper on-premises as opposed to the cloud, so they are pulling them back. Even with the time needed from engineers and the financial losses that occur, they are still finding that it makes financial sense to do so. Several of these examples could have been prevented and valuable dollars saved had these companies performed the right due diligence in the beginning and followed the guidelines we set forth in this book.

Companies, people, and competing vendors are also getting smarter about pricing with the solutions they provide to remain competitive. This is a great thing for the consumer, but it does make it more challenging when evaluating different options for the future of your infrastructure. Even though this book largely serves as an advocate for the cloud, there are going to be situations where it does not make financial sense to migrate certain workloads in the current landscape. While not as common, this does not mean that financial reasons are the only factor though.

You may have other reasons for continuing with your migration. If you follow the principles laid out in this chapter and this book, you should have the understanding you need to be able to make smart decisions on what to do with your workloads. Unfortunately, we will not be able to deliver a definitive direction for you because the answer will always be "it depends."

The next point is that as technical professionals, we are oftentimes the most equipped to be able to answer questions related to cloud costing because we understand the technology the best. And in the cloud, you really need to have an understanding of the technology to determine how the cost translates. It is very difficult otherwise. In an effort for Microsoft to stay true to the "only pay for what you consume" model, this can get complicated to determine. There are some services that are very challenging to figure out with the different variables needed to know that will require technical expertise to answer.

It is also important to realize that evaluating costs is not just something done at the beginning of a project but is something that has to be done throughout the use of it. Notably, every architectural decision that we make in the cloud has a cost tied to it. This means every time you spin up or scale a service, enable a new feature, increase your storage consumption, turn on a backup, and so forth, it comes at a cost. Some costs are fixed and very predictable such as a VM; however, other costs can depend on a lot of factors and really need to be monitored throughout their use. For cost-minded organizations, they expect someone to be able to communicate to them what this looks like over periods of time.

Note   Every architectural decision that we make in the cloud has a cost tied to it.

In an exclusive on-premises environment, it is common to create the technical specifications of what is needed and just send that over to someone who handles procurement and be done with it. Oftentimes, this may occur infrequently since you are provisioning physical hardware for years at a time. As a cloud architect or engineer, you are procurement, and you are making quick decisions all the time that will impact the underlying spend. You are also the one that is designing these solutions and choosing what services you are using, so you must be able to understand what the cost implications of that are. By not understanding this or giving it the proper attention, it could have a serious impact on the budget, as well as the overall status of your project.

As a technical expert, you are also the right person to set the right expectations. There have been a lot of times in our past where we have felt excluded from budget conversations that would have made sense for us to be a part of because we did not have the right amount of expertise. Budget conversations that ultimately impacted our ability to do our job the best we could. This included such things as provisioning hardware or SQL Server licensing. Understanding more about the cost implications pulls you into more of those cost conversations because they are going to want to rely on you and your expertise to be able to come up with solutions. This leads to you and your organization feeling more comfortable about the choices being made.

We fully understand that this is not the most exciting thing, especially when there is all this amazing technology around you. As technologists, technology is what is important to us. But for nontechnical business folks, running a cost-efficient organization is what is important to them. In the end, it does not matter how cool the technology is if your organization refuses to pay for it.

Does the Cloud Save You Money?

A common theme that we often hear through cloud sales and marketing people is that the cloud will save you money. And oftentimes, it is stated very matter-of-factly. But is this true? The answer is that it absolutely can, but not always does. The thought that you can make the decision to just do it and start watching the savings roll in is false and deceptive; however, with proper consideration and planning, it absolutely can. Not every workload and environment is created equal, so it would be tough to make a blanket statement of what will and will not save money, but utilizing the tools and tips throughout this chapter (and even throughout the book) will absolutely assist you in evaluating that decision.

Note   The idea that migrating to the cloud will always save you money is false and deceptive. Although with proper planning and consideration, it can.

As cloud technologies continue to rise in popularity, hardware vendors are also getting smarter and continually finding ways to lower costs to make their offerings more competitive. An example of this is the introduction of hyper-converged infrastructure that bundles together compute, networking, and storage into a single unit to reduce costs. Or restructuring maintenance contracts in a way that influences more frequent hardware purchases is another example. There are several things that they are doing, and it makes sense because if they don't adjust, then they're probably not going to be around very long.

On the caveat though, cloud vendors also realize this, so they are constantly looking at ways to stay competitive on cost as well. Not just among hardware vendors, but also with other big cloud companies with similar offerings. Even though consumer companies may not be comparing different pricing models, we can assure you that the vendors are and working on adjusting accordingly. This is a great thing for the consumer because competition keeps costs lower but also means that the pricing disparity is not always as wide as some would have you believe. This also means that we need to stay on top of what is changing to make the best decision and not just assume that what we have always believed is still true, especially as it relates to pricing.

Getting Started Building a Budget

If you are the one being tasked with creating a budget for your migration, let us be the first to tell you that there is no easy button. It's going to take some effort and diligence to get it right. Microsoft has a complex pricing model that can be difficult to understand that makes it feel like they are constantly nickel-and-diming you. It would be nice if you could just click a button somewhere and it would just tell you what it's going to cost, but unfortunately, it's not that simple, although we are sure some consultants would gladly argue against that. And without a massive overhaul and simplification of cloud costing, we are not sure we ever end up there.

Wouldn't it be simple if the cloud just let you provision resources like you do hardware? For example, you tell it exactly what you want for storage, memory, and CPU without all the extra stuff. While we think it would be doable, the flip side is the costs would likely be much higher than they are now. So instead, we are charged for more miniscule operations that are sometimes difficult to predict because they are based on metrics we have limited insight into or haven't paid attention to before. This leads to confusion and frustration and scares potential suitors because of the unknown.

Ask for Help

When first getting started with any cloud technology service, it can be a daunting task. There are so many options and so much new stuff to learn that you can quickly feel overwhelmed. I can remember a project that I was part of to build out a new enterprise data warehouse in Azure. We were a small team (largely inexperienced in Azure) that was pioneering a large organization's first steps into a cloud infrastructure solution and at the same time transitioning to a myriad of new data services that we had never used before. Factor in all of the new infrastructure work such as governance, security, and networking alongside building a new architecture with services and languages we have never used. And while trying to wrap our head around all of that, we also had to take all of this new information and translate it into a budget with an actionable plan.

With that considered, it's probably not difficult to understand our feelings of being overwhelmed. We were smart enough to know that we couldn't do it alone though and contracted through a third-party vendor and partnered with Microsoft throughout the process. Without the assistance of the outside teams we worked with, this work would have been way more challenging to complete and would have cost us way more in the long run. If you are looking at your project and feeling overwhelmed, we highly recommend reaching out to your partners for assistance or bringing in someone to help.

The natural place to start is by reaching out to your Microsoft representative and letting them know your plans. It is likely they have already been engaged with you about the possibility of moving into their cloud environment, but if not, they will be ecstatic to speak to you. You are utilizing their product after all and spending your money with them, so they have every incentive to want to see you succeed.

Getting started and early success are crucial to the overall project. We often envision this process as having a rowboat on a deserted beach you are trying to get off of. The hardest part to accomplish is getting your boat over the tide. Continued failed attempts to make it over the tide and you may just feel you can stay on the beach and make the best of the situation. Press through and you are well on your way. You may still hit choppy waters along the way, but the hard part is completed.

The metaphorical tide can be different based on your organization; sometimes, it could be a technical or security challenge, but it also could be a budgeting or cost challenge. It is for whatever questions you need to have answered, but you are ready to be all in. In our experience, this is a pivotal place that Microsoft can and is willing to assist. Ultimately, if you cannot get the groundwork laid and get started, then the project is likely doomed.

We are not Microsoft employees, so we cannot speak for them about all of their offerings – that is something you will need to discuss with them. However, we can share that through our experiences, they have provided us access to expert resources, provided documentation, evaluated decisions, and connected us with outside resources that could help. Without their involvement in the projects we have been part of, we are not sure how successful we would have been for several of them. In a lot of ways, they became an extension of your team.

Another option is also using a reputable third party that specializes in cloud migration to assist you. Yes, this will come with a cost, but it can be money well spent if you are projecting this to be the future of the organization. As mentioned, we believe one of the most important things is creating a solid foundation that you can build on going forward. We believe this is one of the most important things you can invest in early on and should at least be considered.

There are tons of tools and documentation available that can guide you through any decision, but with all the newness, it's easy to get overwhelmed. If you are, know that you are not alone in that feeling. Work with your existing partners to assist you, and if the load is too heavy, then we would encourage you to consider bringing in additional experts.

Introducing the FLAT Method

Before we start diving into the details of building a budget, there are several unique considerations or decisions that you need to factor before and during the budget building process. We created a method that is easy to remember called the FLAT method that we believe sums up these important considerations.

The FLAT method is made up of four components with each letter representing a different decision/consideration:

- Familiarity

- Location

- Architecture

- Translation

Each one of these will play an important role in building out our budget, and we will dive into each one of them in more detail in the sections to come.

Familiarity

One of the most important things you can do when getting started with any cloud technology is to do as much as you can to get familiar with it. It is very hard to budget for and speak to something you have no experience with. For us and several other technical professionals, we often learn best from getting hands on with something. Videos, training material, and blog posts are great, but it is a different experience when you are the one steering the ship. And after all, if your budget is approved, then you are going to need to be able to understand and use it anyways.

What we often encourage people to do who are new to Azure is talk to their company to see if a designated amount of time can be set aside to be able to just get some familiarity with the solution. This is to allow the team to go in, click around, follow tutorials, explore features, and possibly even make a few mistakes along the way that can be learning opportunities.

This is important from a usability standpoint but also on being able to see cost outcomes of choices you make. Use that time to observe what the cost is for certain actions so you can tell how dollars are being allocated. You will be able to see firsthand where most of the cost is consumed so you can focus on those areas as well as see places where cost is fractional and possibly not as much of a concern.

Oftentimes, this stage of the project is going to come with a cost unless you are able to take advantage of any existing programs. We encourage you to have conversations with your Microsoft representative to discuss the options here if you are unsure if there are any free credit options available.

If you have a Visual Studio subscription, then you already have free Azure credits that are allocated to you each month that can be used. This is a little bit of hidden benefit that we have found that several developers fail to take advantage of and can range from $50 to $150 based on your Visual Studio subscription level. To see your number of credits, just log in to your Visual Studio subscription online and navigate to the benefits area to be able to activate. This program also provides several other benefits for developers, so if you are not part of this already, we would recommend looking into this.

If your company is not interested in doing that, then there are other options that are also available. First, you can sign up for Visual Studio Dev Essentials (`https://visualstudio.microsoft.com/dev-essentials/`), which is a free program that provides you access to several developer tools, as well as a ton of valuable information. You may be already familiar with this great program and may have used it already for other aspects of your work, but you may not realize that there is a free Azure credit that is also part of it. As of this writing, the credit is for $200 along with a 12-month free service to experiment in your own personal dev\test sandbox. This may not be something that your company advocates for, but it is something that you can do personally.

And as a last resort, you can always request a small budget to be set aside within your organization or create a personal Azure account and utilize the free credits available within that. A personal account will require a credit card to sign up, but it will not charge the card unless you explicitly allow it to.

For doing things such as building familiarity, the spend should stay relatively low if you use common sense with what you are doing. For example, without knowing anything about pricing, we know that a 64-core virtual machine is going to cost more than a 2-core virtual machine, so you are not going to do that. And why would you? If you are just getting familiar with the services, then you can spin things up at the lowest tiers possible. The functionality is often the same or similar at the lower tier, just less resources. So, unless you are running specific tests that require higher tiers, you should be able to keep costs low. If you are setting aside dollars to be able to do it, we generally suggest around 100–150 dollars per person.

While we have mentioned several programs and there may be several others available not mentioned here, we would highly recommend finding a way to start getting familiar with the functionality and features of Azure. The more familiarity you can get, the more equipped you will be to build a budget.

No one with reasonable expectations is going to expect you to know everything at this stage. You are going to consistently be learning as you go through the process even when you start your first implementation. Familiarity is not just a pretask but also a during task. This is why one of the best pieces of advice we can give when starting a cloud migration is to start small.

Start with smaller, less critical workloads that give your team an opportunity to adapt and familiarize. Several of the cloud migration failures we have seen were the result of trying to do too much too quickly. If your team is not experienced with the cloud and you try to shift your entire infrastructure, this is going to pose challenges. Yes, it may work, but it's probably going to come at the cost of overspending on your Azure run rate and spending significant dollars to do the work and maintain it.

Tip   One of the best pieces of advice we can give on starting a cloud migration is to start small. Most unsuccessful migrations are the result of trying to do too much too fast.

Location

Another important thing to think about when getting started is what region you are going to be running in. A *region* is a set of data centers in a region of the world where your services are going to be hosted. The choices that you make here could have an impact on the effectiveness of your overall solution, as well as an impact on the overall cost.

There are some important considerations that need to be reviewed when making the decision of which region to build in. A lot of times this is easy for people to overlook because the expectation is that you are best off with the region geographically located closest to you. It makes a lot of sense to think about it that way based on an understanding of how networks work, but it's actually not always the best option, though we will concede that it often can be.

Here is a list of the considerations that we suggest factoring in when making this decision:

- Being sure the region supports the service you are planning to use

- Evaluating latency

- Evaluating cost

You might be surprised to find out that not all services that are part of Azure exist in every region, and this is especially true for newer ones. So as you consider what services make up your architecture (covered in the next section), you will want to **be sure the region supports that service**. Yes, you can run services in multiple regions, and sometimes that may be required; however, if the said service needs to be able to frequently communicate with others, then you could be introducing added latency. It might be better to choose a region that has everything you need rather than spreading them out.

The next component and probably the most essential is **evaluating the latency**. The reason people choose the closest region to them geographically is because they anticipate latency being the lowest. While often that is correct, we have seen situations where this has not been the case. We recommend taking the top three regions closest to you and running some tests to be able to check the latency. Two of our favorite tools to use for this are *Azurespeed* (`www.azurespeed.com/`) and *Azurespeedtest* (`https://azurespeedtest.azurewebsites.net/`). Azurespeed shows network latency in real time from your IP location, and Azurespeedtest measures your latency to BLOB storage. These are really simple web-based tests that can be done to check your latency.

In addition to evaluating latency, you also need to consider where your application servers are hosted or where your end users will be accessing the data, not necessarily where you are located physically. An example of this would be if you worked out of a corporate office in Nashville, Tennessee, but your corporate data center and entry point for any end users was located in San Francisco, California. You would not want to choose a region that is closer to Nashville; you would instead consider choosing a region that is closer to San Francisco because that is where our users are going to experience the lowest latency. Sometimes, you have multiple entry points and users spread out all over various regions that could warrant running in multiple regions. In those cases, you are going to want to really think through how you structure that.

The last component and what makes all of this relevant to this chapter is **evaluating the cost**. The cost of a service in Azure is not static and can change between regions. For example, if we have a D4 Azure Virtual Machine (4 cores, 16 GB RAM, excluding storage) in the East US region running 24×7, it would currently cost $566.53, but if you build the same virtual machine in the West US region, the cost raises to $589.89. It's the same build instance in Azure, but the cost of it is different between regions.

This is also not just specific to a single service in Azure but can apply to all. In our experience, the cost differences have been small, but this may not always be the case and can certainly be impacted by the volume of things you will have running the cloud. In addition to cost disparity, you will also want to consider egress charges which we will cover in greater detail later in the chapter.

Latency is likely going to be one of the most important considerations when determining a region, but if you are experiencing comparable performance between multiple ones, it would be worthwhile to evaluate the cost differences as an added element in your decision-making process.

While location may not be the most prominent factor in your budget-building process, it is one that is worth evaluating in an effort to reduce your costs.

Architecture

Another important aspect of building a budget is determining the different options that would work for your architecture. In this instance, architecture relates to what services or combination thereof you are going to use in the cloud to run your workload. One of the great (and sometimes confusing) aspects of the cloud is there are several different ways to be able to do things. For example, if you want to move a SQL database to the cloud, you can use an Azure VM running SQL, Azure SQL Database, Managed Instance, or a Synapse pool.

Going a step further, we have even more additional options with each of these that can fundamentally change what you pay for and how you use the service. Such as with Azure SQL Database, you can utilize serverless, hyperscale, or elastic pools instead of a standard database. Each of these would come with different benefits and may work better in certain use case scenarios. We recommend checking out the planning for your migration portion of the book to get a better understanding of the various architectures.

It is impossible to separate architecture from cost management if your goal is to keep costs low. You need to either be able to translate your architecture into a cost that makes sense or translate your cost into an architecture that makes sense. This is going to require some understanding on both parts. For budgeting, you need to have an idea of what services you can use to evaluate the options. You do not have to know the final answer yet, because the deciding factor could come down to which design is more cost-effective. On many of the projects we have worked on, this has often been the case. We were often tasked with finding the solution that best fits our needs, but for the lowest cost.

Note   It is impossible to separate architecture from cost management if your goal is to keep costs low.

Familiarity also becomes an important part of building your Azure architecture because you need to understand the features of each service and its limitations to understand what works best for you. Cost is an important aspect for most, but knowing the solution will work is essential. One of the worst things you can do is build a budget for something that gets approved and later find out it won't work and instead discover you needed a more expensive solution. If you are ever unsure, run through test scenarios to make sure it works as you anticipate it to be ahead of submitting a budget for approval.

Translation

When budgeting for a migration, you must be able to take existing on-premises workload and *translate* those to the cloud. By translation, we are referring to resources such as compute, memory, disks, and their subsequent performance. You could match the resources 1:1, but this is often not a good idea to do blindly since it can easily lead to wasted spending. Rather, it is a good idea to evaluate those workloads, so you only build for what is needed.

There are several tools that can assist you in the effort. We are listing several common examples, but this list could easily include others. You can also read the section of the book on performing a migration to learn more about these tools.

- Database Migration Assistant (DMA)

- Database Experimentation Assistant (DEA)

- Azure Migrate

- Data Migration Guide

- DTU Calculator

- Monitoring tools

- Third-party tools

When you adopt a cloud technology, your mindset does have to change from how you manage on-premises because in the cloud, you are paying for what is allocated if you do not want to overallocate more resources than what is absolutely needed. This is especially true for compute costs, which are oftentimes the most expensive resources in the cloud.

When we were Database Administrators, we had a scale we would often use to look at our on-premises database servers. Within this scale, we liked to try to keep our production SQL Servers averaging around 30% CPU utilization. If it ran higher than that, then it was something the team would investigate and try to remediate. Sometimes, it was resolved with some performance tuning, but if it was workload volume related, we might have looked at opportunities to increase the resources.

Within the cloud, this mindset of low thresholds does not make as much sense anymore because it would lead to overspending. Instead, we want our workloads to consume most of the resources allocated while not impeding our workload. We typically try to run our cloud workloads at near 80% utilization as a starting point because we do not want to be paying for something we are not using. If for some reason it's not enough, then we can scale up. Or if it's only busy for a couple hours of the day, we can automate it to run at the higher tier for those two hours and lower it the remaning time to limit paying for the higher tier computer.

Remember that on-premises workloads are built to handle workloads for longer periods, whereas cloud workloads are built to handle the workload for the moment.

Total Cost of Ownership Calculator

The Total Cost of Ownership (TCO) Calculator (`https://azure.microsoft.com/en-us/pricing/tco/calculator/`) is a tool that can be used to estimate the cost savings from migrating your workloads to Azure. It does this by allowing you to build a comparison using very detailed metrics from an on-premises workload and translate that to the cloud. You simply input the values describing your workload, and at the end, you get informative easy-to-read charts that can be used to evaluate your decision.

As you are probably aware, there are a lot more costs tied to on-premises workload than just servers and licenses. While these are absolutely factors, there are other things that are often easy to overlook such as electricity, storage, labor, data center costs, and more.

Figure 3-1 shows a workload priced out in the calculator. This calculator takes all that into consideration to evaluate whether you can save money or not using detailed data about those things. While the tool is very useful, it's not one we generally use for every migration because it can be time-consuming to get right and requires a lot of information gathering. However, if comparing your existing spend in totality versus your total spend in Azure to build an ROI is important, then this is definitely a tool that can assist you.

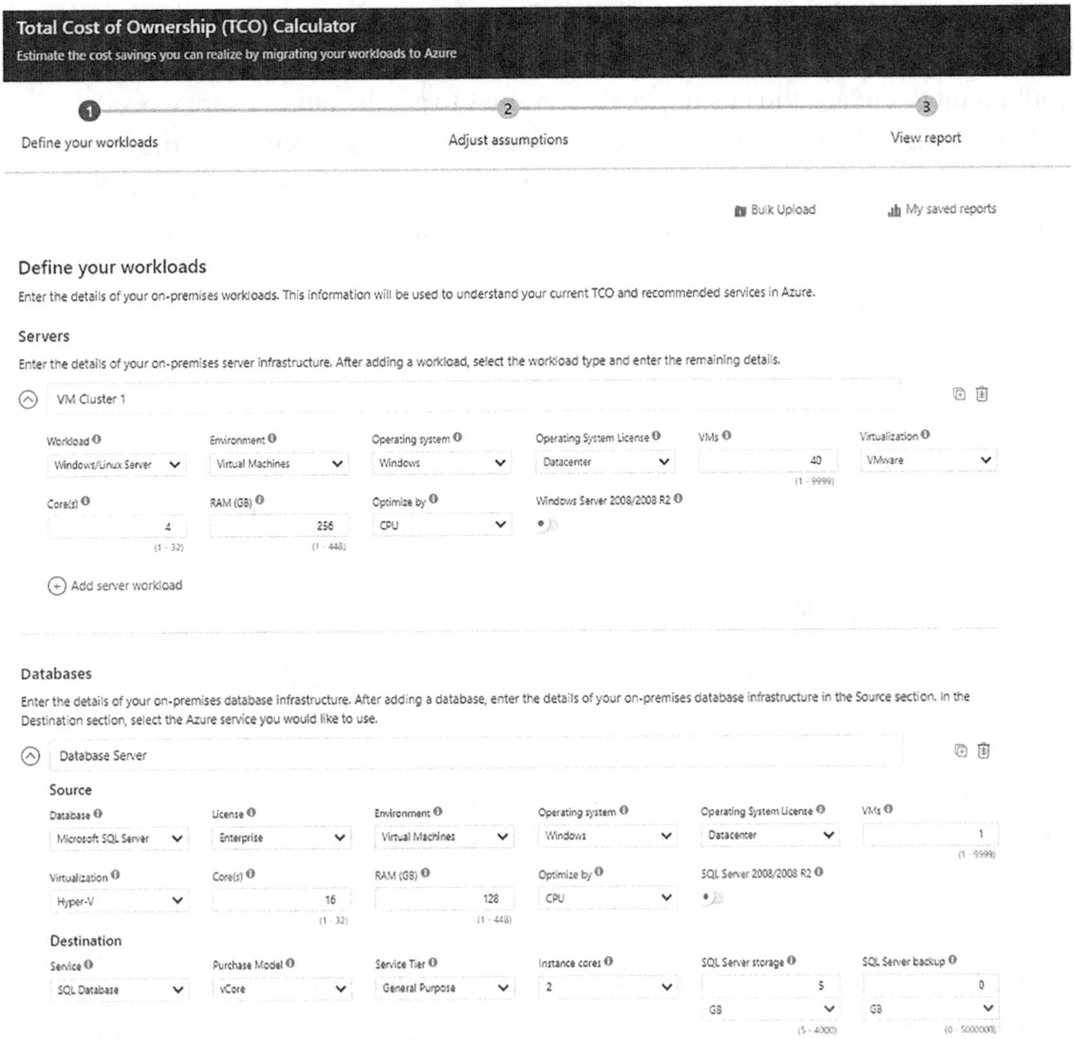

Figure 3-1. *Examples of defining your workloads in the TCO calculator*

Within the calculator, there are three primary components: *Define your workloads*, *Adjust assumptions*, and *View Report*. Defining your workload is where you will enter information about your on-premises workload. Here, you can add details on servers, databases, storage, and networking. The next part is about adjusting assumptions on how much you are being charged for various components such as hardware, labor, electricity, data center costs, and so forth.

There are default values that are added that are industry averages accredited by Nucleus Research. The default values can be a great quick look; however, you would want to update and customize those values to get the most accurate results as they can significantly vary from your situation. The last part is viewing the report (see Figure 3-2), which takes the workload information and assumptions from the calculator, creates a comparable build-out in Azure, and compares the results in the form of several charts.

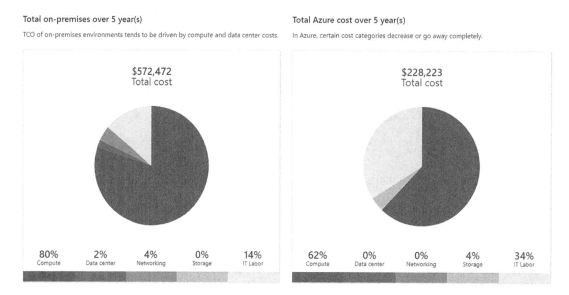

Figure 3-2. *Example of report comparisons produced in the TCO calculator*

It seems like it would be easy to always skew the results so that the cloud always looks like the more cost-effective solution, but we have gone through scenarios with the TCO calculator that actually provided the opposite. With that, we feel good for the most part with the results provided if our assumptions are correct. Additionally, we have noted that the estimated cloud cost often runs on the higher end, so if you follow some of the guidelines in this book, you may be able to get the cost lower.

Azure Pricing Calculator

The Azure Pricing Calculator is the most important tool in your toolset when it comes to building a budget. It is a free web-based tool (`https://azure.microsoft.com/en-us/pricing/calculator/`) Microsoft provides that can be used to build estimates for your cloud migration. This is the place where all the ideas come together and you start building your budget.

Note   The Azure Pricing Calculator is the most important tool in your toolset when it comes to building a budget.

Whenever you first access this page, we highly recommend that you first log in using your existing Azure account or company account if you are starting new. This will not only allow you to save progress, but it will also allow you to see specific pricing for your organization. If you have an Enterprise Agreement (EA) with Microsoft, those discounts would be reflected here.

The tool is very simple to navigate and is made up of just a few core components that you see in Figure 3-3. The first is the products screen. Here, you have access to all of the services that exist within the Azure library. You can simply type in the name of the service you are looking for or navigate through the types of services to find the one you are looking for. Whenever you find a service you want to add to your estimate, you just click on it to add it. If you already have an estimate created, it will add the service to it, and if not, it will create a new one for you.

Figure 3-3. *Selecting products in the Azure Pricing Calculator*

The next core component is the estimate itself, shown in Figure 3-4. Here, you can see the products that you previously added along with more detailed configurable options for them. When the product gets added to an estimate, you unlock the ability to configure them to match your build requirements to ultimately unlock your price. The pricing is calculated based on how Microsoft will bill you for the service and uses real-time pricing information. These options are not exclusive to the build though, and you will have added metrics that you will need to fill in related to the workload itself. These can sometimes be things as simple as the number of reads, writes, or executions.

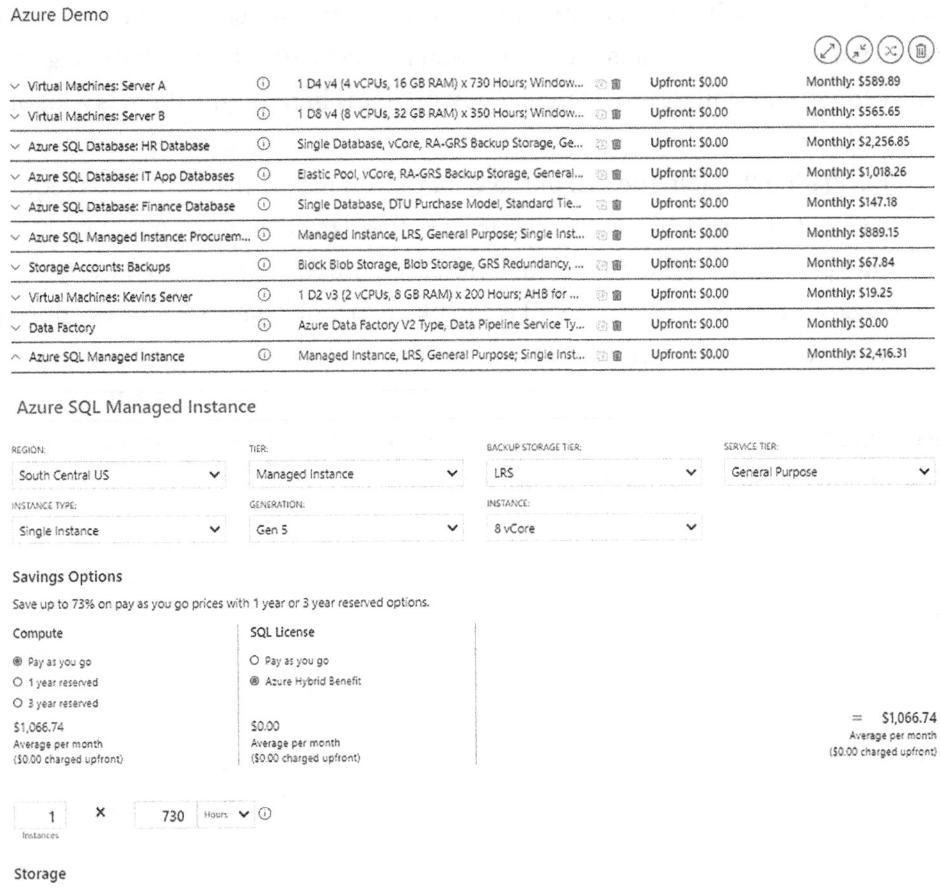

Azure Demo

∨ Virtual Machines: Server A	ⓘ	1 D4 v4 (4 vCPUs, 16 GB RAM) x 730 Hours; Window... 🗐 🗑	Upfront: $0.00	Monthly: $589.89
∨ Virtual Machines: Server B	ⓘ	1 D8 v4 (8 vCPUs, 32 GB RAM) x 350 Hours; Window... 🗐 🗑	Upfront: $0.00	Monthly: $565.65
∨ Azure SQL Database: HR Database	ⓘ	Single Database, vCore, RA-GRS Backup Storage, Ge... 🗐 🗑	Upfront: $0.00	Monthly: $2,256.85
∨ Azure SQL Database: IT App Databases	ⓘ	Elastic Pool, vCore, RA-GRS Backup Storage, General... 🗐 🗑	Upfront: $0.00	Monthly: $1,018.26
∨ Azure SQL Database: Finance Database	ⓘ	Single Database, DTU Purchase Model, Standard Tie... 🗐 🗑	Upfront: $0.00	Monthly: $147.18
∨ Azure SQL Managed Instance: Procurem...	ⓘ	Managed Instance, LRS, General Purpose; Single Inst... 🗐 🗑	Upfront: $0.00	Monthly: $889.15
∨ Storage Accounts: Backups	ⓘ	Block Blob Storage, Blob Storage, GRS Redundancy, ... 🗐 🗑	Upfront: $0.00	Monthly: $67.84
∨ Virtual Machines: Kevins Server	ⓘ	1 D2 v3 (2 vCPUs, 8 GB RAM) x 200 Hours; AHB for ... 🗐 🗑	Upfront: $0.00	Monthly: $19.25
∨ Data Factory	ⓘ	Azure Data Factory V2 Type, Data Pipeline Service Ty... 🗐 🗑	Upfront: $0.00	Monthly: $0.00
∧ Azure SQL Managed Instance	ⓘ	Managed Instance, LRS, General Purpose; Single Inst... 🗐 🗑	Upfront: $0.00	Monthly: $2,416.31

Azure SQL Managed Instance

REGION:	TIER:	BACKUP STORAGE TIER:	SERVICE TIER:
South Central US ∨	Managed Instance ∨	LRS ∨	General Purpose ∨

INSTANCE TYPE:	GENERATION:	INSTANCE:
Single Instance ∨	Gen 5 ∨	8 vCore ∨

Savings Options

Save up to 73% on pay as you go prices with 1 year or 3 year reserved options.

Compute	SQL License	
◉ Pay as you go	○ Pay as you go	
○ 1 year reserved	◉ Azure Hybrid Benefit	
○ 3 year reserved		
$1,066.74	$0.00	= $1,066.74
Average per month	Average per month	Average per month
($0.00 charged upfront)	($0.00 charged upfront)	($0.00 charged upfront)

[1] × [730] Hours ∨ ⓘ
Instances

Storage

Figure 3-4. *Example estimate created in the Azure Pricing Calculator*

You also have other configurable options within the calculator you will need to define that are not directly related to workload. The first of these is the region which we discussed earlier in the chapter. At the point you are ready to build an estimate, you should at least have an idea of which regions you may want to use. The next options are related to cost-saving options, most notably Azure Reservations, which is a commit-to-spend program, and Azure Hybrid Benefit, which allows you to transfer Windows and SQL Server licenses you already own to the cloud. We will discuss both of these in more detail later in the chapter, but we want to highlight that both of these are part of the calculator.

After adding a few products and getting started on your estimate, you quickly realize that pricing is not as straightforward as you might expect and it's easy to feel overwhelmed. We can certainly attest to the sentiment. The pricing seems like it's very much a model that is built to nickel and dime you or even confuse you so much you just give up, but we are not sure that is the intent even if it feels that way. With managing large data centers with millions of concurrent workloads happening, there is a lot of complexity.

When navigating through the pricing calculator, you may run across items requiring a value that you are not sure about. In some cases, there may be information in the calculator that provides more context, but if not, then the Azure Pricing Guide (`https://azure.microsoft.com/en-us/pricing/`) is the place to look. Figure 3-5 shows this guide as it appears in your browser. Here, you can view detailed information for the pricing around any service in the Azure library. Using a similar search experience that the Pricing Calculator uses, you can just search the service you want to know more about and click on it to reveal more information.

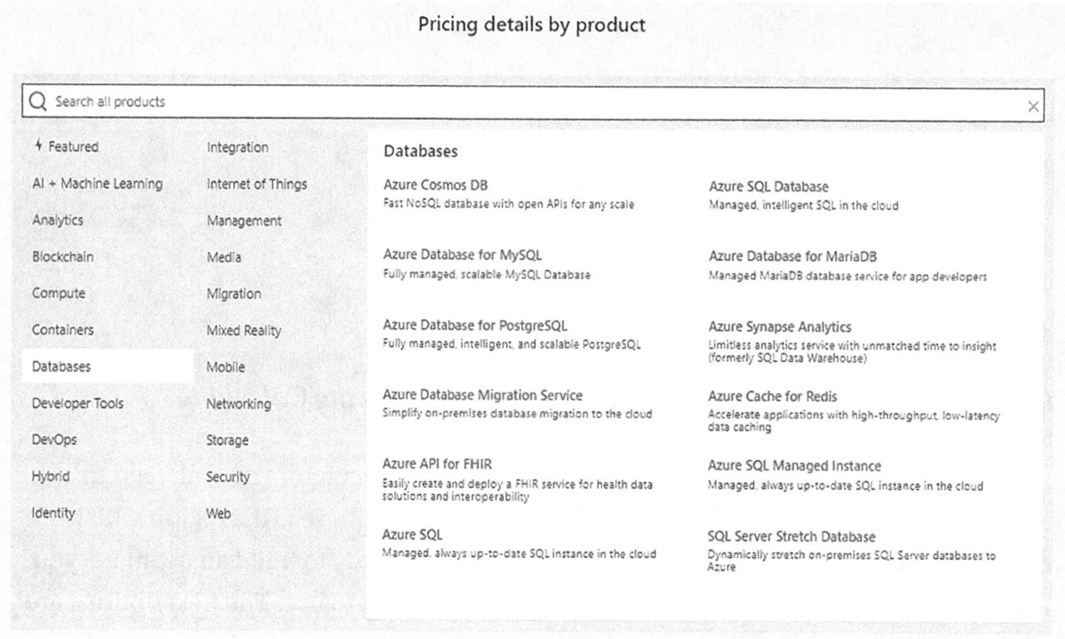

Figure 3-5. *The Azure Pricing Guide product page*

Outside of just being able to build estimates, the Azure Pricing Calculator also has a variety of other features to make this experience more user-friendly. An important feature is the ability to save an estimate and return to it at a later time. This is great because it allows you to work off the same estimate throughout the analysis process as you learn. You also have the ability to be able to export estimates to share with partners or members of your team. The export is what we ultimately use as our budget for our Azure run rate and to keep records over periods of time.

As we reviewed earlier, the tool is simple to start using. The difficult part with it is understanding what values need to be entered. This is where the information we shared with the FLAT method becomes very relevant. Familiarity, Location, Architecture, and Translation are essential components when building an estimate.

Important Calculator Considerations

As you can see, the Azure calculator will be the most important tool that you will use throughout the budgeting process; however, there are some important considerations when using this tool that are not explicitly defined.

When inputting values into the calculator, try not to underestimate or guess and make sure what you are entering is backed up with data. Since the output of what you enter into the calculator is ultimately going to be your budget, you want it to be as close as you can. Utilize the tools mentioned earlier in the "Translation" section to assist you in defining the comparable workload in Azure. Depending on what you are moving, you may also want to go through some test exercises where you build out the service in Azure and evaluate performance to make sure it works as expected. Whatever you do, you want to make sure that the values you input make sense to you and you are comfortable with them.

The calculator does not tell you which products have dependencies on others. This can lead to missing items in your estimate because you weren't aware that you need another service to make it work. This is a difficult one to overcome especially if you are new to Azure. Testing can be a way to identify these examples, as well as following our advice to start slow with a migration.

Always hit the save button when you are making changes. Whenever you make changes to your estimate, be sure to remember to hit the save button. There is no autosave, so if you close the window without explicitly saving, then your changes will be lost. We have personally made this mistake several times, and it's heartbreaking to spend a day inputting values to see it all wiped away.

Egress is not factored. Your estimate is not going to factor egress of data, or data leaving Azure, into the cost. Depending on your workload, this could really impact your underlying budget. Later on in this chapter, we will look at this more in depth.

The calculator is meant to be used throughout the analysis process and beyond. Being able to assess workloads and build a budget is something that will take some time. As your project evolves, you continue to learn, and your architecture changes, you should continually be updating it. Even once you successfully migrate the workload, this estimate still has a place in cost management. In the next chapter, we will talk about how you use this budget to create automated actions in Azure.

The data is exportable. This is an important feature because it allows you not only to export your build for tracking purposes but also to create copies that can be shared with your team and your partners. The Azure calculator does not track versions of a budget, only the latest, so if you want to track your estimated cost over time, you will need to export.

Getting the Most from the Calculator

A simple method that we like to use when using the calculator is the calculate, evaluate, and infiltrate method, which sums up a lot of the principles we have already looked at. Hopefully, this provides an easy-to-use way of thinking about the calculator.

The first of these is **calculate**. Within this stage, it's all about creating the estimates and experimenting. It's taking the ideas you have inputting them to see what the cost output is. This is the area where you can try out different architectures and services.

The next one is **evaluate**. This is all about reviewing your design and constantly asking yourself questions throughout the process. These are questions about the feasibility and business impact to your design, such as follows: What will this change impact? Are there different approaches you can take to this? Are there optimization opportunities? What value does this add to the organization?

The next stage is **infiltrate**. With this one, it's about understanding the product the best you can. If there are uncertainties about functionality or cost impacts, then you should spin up the service in a sandbox and test. It's important to understand the options, the limits, and the product the best you can.

This process is not intended to be a one and done, but it is intended to be cycled through as much as needed to be comfortable with the output. As you experiment with different designs, you may evaluate and infiltrate it before starting over with a different one and then compare the results. There are a lot of ways that you can look at utilizing this, but this is a simple and easy-to-remember way of doing it.

Big Scary Egress

There are two different terms for data going into the cloud and data coming out. *Ingress* refers to data going into the cloud, which is free, and *egress* refers to data leaving the cloud, which is not free. One of the scariest things for organizations thinking of moving to the cloud is egress because it is often difficult to predict and is accompanied by costs not factored into most of the tools. Moreover, egress charges will differ depending on the type of network transfer. For example, network egress between regions within North America is less expensive on a per-GB basis than transfers between regions in Asia.

Careful consideration does need to be given to workloads that are being migrated to review how much data will be leaving the cloud. It is important to consider how you will be serving up data to its destination, whether it's through extracts, directly to the end users, application or web servers, or any other means. Will the data stay within the cloud or will the data be leaving it? If it stays within the cloud, then you do not incur egress charges. But if your data needs to leave the cloud, then you will.

This is why oftentimes if you are moving a database to the cloud, you should consider moving all of the collective components with it to keep it contained together. For example, it may not be a good idea to migrate the database for a highly transactional system and leave the application running on-premises. If it is required to keep components separated, then it may not be a good candidate for a cloud migration.

Tip When migrating a data workload to Azure, consider migrating anything with it that can reduce the amount of data leaving Azure. A good example is application and web servers.

Egress is also critical when considering options for High Availability and Disaster Recovery (HA/DR) as data sent between AlwaysOn Availability Group Nodes in different availability zones or regions may incur substantial costs.

With any migration, egress is unavoidable and should be expected. There is always going to be a need to pull data out. The trick is avoiding doing it unnecessarily and understanding when it will occur. While you do pay for it, it may not be as bad as you might think. As of this writing, the current cost of egress starts around $0.087 per GB for North America and Europe, which is around $88 per TB and goes down from there.

For other parts of the world though, it can be as high as $0.18 per GB. We recommend checking the latest rates to be sure if this is a concern. Additionally, if you know this might be an issue, you can look into implementing an ExpressRoute, which offers reduced outbound charges or paying a flat rate for outbound data. This can come with a big price tag, but you may find it worth it.

If egress is a concern and you think it may balloon and impact the budget, then start with migrating smaller workloads and evaluate its impacts and gradually increase. With smaller workloads, you can easily and most cost-efficiently pull them back if the outcome is not desired. If you start with migrating a lot of workloads with large data volumes, then your rollback plan becomes a lot more costly.

Networking Cost Considerations

The cloud is only as useful as your ability to use it, and this starts with being able to connect to it. This involves building a secure gateway between the users and your Azure environment, and as you have probably already guessed, this is going to come with some costs. Commonly, you will be using either a Virtual Network (VNET) that creates a private network in the cloud or a VPN Gateway that can be used to connect an on-premises environment to your virtual networks in Azure.

A Virtual Network is free in Azure; however, many services that you use within it are not. These include public and reserved IP addresses, which is usually a nominal cost, and any Application or VPN Gateways you create within them. There is also a component called VNET Peering that allows you to link virtual networks, and for these, you will incur bandwidth charges. This includes both inbound and outbound. For a VPN Gateway, you will incur costs for it for the amount of time the gateway is available. There are several different types of gateways that can be created, and we recommend checking out the official documentation to understand these better. In our experience, charges related to Virtual Networks or Gateways have been some of our lower costs. The exception to this would be if you have a lot of data leaving Azure, thus resulting in high bandwidth charges in the form of egress.

Outside of this, there is another option that you may consider for your project, and that is using an ExpressRoute. An *ExpressRoute* is used to create a private connection from your on-premises or colocation environment directly into the Azure data centers. Because these are private connections, you do not traverse the public Internet, making them more reliable, faster, and more secure. These are great investments if you know if you are going to have constant communication from your on-premises environment to Azure either direction. If your workload is exclusive to the cloud and the communication is nonexistent or minimal, you may be fine without this. These can get really expensive depending on your bandwidth needs, so you will want to check into this before making a final decision.

When you look at this in the Azure Pricing Calculator, you actually do not get the full impact of the cost. That is because the Azure cost is a small piece of the overall spend. Most of the cost will come from your telecom service provider to cover the cost and implementation of the connection. These costs can add up to several thousand in implementation fees and several thousand on a monthly basis. If this is something you are considering, you are going to want to have some conversations early on to be sure this becomes part of your budgeted expenses.

Reducing Your Azure Costs

Now that you understand the tools and some of the considerations needed to build a budget, we can start looking at programs and techniques that can be used to reduce your costs even further. These can result in significant savings for your overall project and could play a large role in determining whether or not the cloud is a cheaper option or not. And maybe more importantly, it could be the deciding factor on whether your project is approved to proceed or not.

Azure Hybrid Benefit and "Bring Your Own License"

Azure Hybrid Benefit (AHB) is a benefit that allows you to take your already-owned Windows and SQL Server licenses with you to the cloud. As a result, you can save some big dollars on the services you use. The alternative to this is pay as you go, which means the licensing cost is baked into the product and you will pay the full amount.

As you can probably guess, there are some limitations to using this. We think it is best to look at these constraints first, so you have an idea whether you qualify:

- You MUST have software assurance on your licenses, and it must be maintained. You need to have it now, and you must keep it, and if at any point you drop software assurance on those licenses, then you do lose this benefit.

- This option is exclusive to Azure. There are options to continue to use your licenses with other cloud providers as well, so all is not lost, but you need to understand their terms. Before proceeding, I suggest checking out the official terms with the cloud provider you are considering.

- It cannot be applied retroactively. Basically, if you have a service already running that qualifies for this but did not apply for it, you are not entitled to any refund or adjustment. The savings only start at the time you implement it.

The main and most notable benefit is it allows you to transfer your licenses to the cloud, saving you money, which is awesome, but there are some other benefits it also provides:

- Security updates for recently retired versions of SQL Server and Windows Server. If you are stuck on this version on-prem, then you are out of support unless you paid for the very expensive extended support. Moving to an Azure VM in the cloud provides you an opportunity to work through getting upgraded without having a security risk.

- It provides you with a 180-day grace period to complete your migration. This means that you can continue to use the same license on your on-prem server and cloud service at the same time. Note, though, that you can only do this for purposes of the migration, so no sneaky stuff here.

- Combined with reserved pricing, you can really get some whopping savings. Reserved pricing is you making a commitment for service use for a designated period of time. The more years you commit, the higher the savings.

You may be wondering how much you can really save, and Azure has a calculator (Figure 3-6) that can help you determine that. The Azure calculator is the primary example that we looked at earlier in the chapter, but there is also an Azure Hybrid Benefit Savings calculator (https://azure.microsoft.com/en-us/pricing/hybrid-benefit/#calculator). With this calculator, you can plug in your build and see your savings; it is specific to calculating Azure Hybrid Benefit savings.

Azure Hybrid Benefit Savings Calculator

Windows Server VMs	SQL Server VMs	SQL Database

Enter the number of SQL Server license cores with Software Assurance

Standard Edition
1

Enterprise Edition
2

Enter planned Azure deployment of SQL Server Instances

Region
Central US

Type:
Managed Instance

Tier:
General Purpose

Generation:
Gen 4

Instance
8 cores $2.262/hour

Hours / month
730

Eligible number of Managed Instance General Purpose instances based on your Instance selection

1

Monthly Estimates

Without Azure Hybrid Benefit per month	$1,650.542
With Azure Hybrid Benefit per month	$1,066.741
Savings across eligible databases per month	**$583.802**
	(35.4% savings)

Annual Estimates

Your estimated annual savings on Azure across all databases	**$7,005.618**

You have a 180-day grace period to use licenses both on-premises and in the cloud to facilitate migration. This calculator is to help estimate savings range when using the Azure Hybrid Benefit for SQL Server licenses that include Software Assurance. Your actual savings may vary.

Figure 3-6. *The Azure Hybrid Benefit Savings calculator*

The same rules that exist on-prem for SQL and Windows licensing also exist in the cloud unless noted.

One notable gotcha that is easy to overlook for SQL Server is the licensing minimum, which is four cores (two licenses). You can build services that use less cores than this; however, keep in mind that you will pay for core licenses you are not using. If you need less, consider using elastic pools, provisioning by DTU with Azure SQL Database, or some other combination method.

In terms of how your SQL licenses translate, here are a few considerations to be aware of:

- 1 Enterprise Edition license (2 cores) = 4 Standard Edition licenses (8 cores) for general-purpose service tiers (×4 multiplier). Standard licenses are 1:1.

- MSRP Standard Edition licenses are 26% of the total cost of Enterprise, so Microsoft is giving you a little bit of bargain that can save you a few dollars here should you choose to transfer these.

- 1 Enterprise license (2 cores) = 1 Enterprise license (2 cores). You must use Enterprise edition for anything that uses the Business Critical service tier.

So how do you activate this feature? Whenever you go to build an eligible service in Azure, just check the box yes when it asks if you have already had licenses that you own that you plan to use.

If you are thinking of moving to Azure and you do not have software assurance or licenses available to be transferred, should you buy the license or use the pay-as-you-go option? Spoiler alert, it probably depends.

Let's look at an example Azure SQL Database with the following specs:

4 Cores, Gen 5, South Central Region, General Purpose Service Tier, Standard Edition

With this example and using the Azure Hybrid Benefit calculator, we see that we can accrue a savings today of $292 a month using Azure Hybrid Benefit (or a 35.4% overall savings). Or another way of looking at this is without Azure Hybrid Benefit, we are paying $292 a month more for pay-as-you-go licensing. Looking at an entire year, that would be $3,504, so let us see how this compares to buying the license instead.

Because we are using MSRP Azure pricing, we will do the same with the SQL Server Standard Edition licenses. Today, if we buy two Standard Edition licenses (which is what we will need for this scenario), it will cost us $7,436. This is the cost to get just the license though and does not include our cost for software assurance, which will be needed to use this in Azure. As a starting point, we recommend factoring in 30% of the total cost for this, which translates to roughly $2,231 per year. So your first-year investment would be around $9,667 and then an additional $2,231 each year after to maintain SA. Figure 3-7 shows year-over-year total investment using both methods.

	Pay as you go	Buy your own
Year 1	$3,504	$9,667
Year 2	$7,008	$11,898
Year 3	$10,512	$14,129
Year 4	$14,016	$16,360
Year 5	$17,520	$18,591
Year 6	$21,024	$20,822

Figure 3-7. *Diagram showing the cost difference between buying SQL licenses outright or paying for them as part of the service using estimated costs*

Using these numbers, it would take you six years before buying your own license would start to save you money. And Azure changes so often that even by that point, so many factors could have changed that pay as you go still ends up the cheaper option. Looking at it another way, it would take over two years on the pay-as-you-go license before you meet the initial cost of the Standard Edition licenses (excluding SA) as a basis if you are paying less. Before deciding here though, we would recommend factoring in your own licensing numbers to do a comparison since it can vary based on agreements.

Also, in addition to cost factors, you may want to consider purchasing your own license if you think there is a chance you may need to reuse it down the road such as migrating out of the cloud or if you are using the cloud as a temporary stop gap for a broader project. With a pay-as-you-go option, the license is baked into the service you are running, so you cannot use that for anything else.

Another consideration is that whenever you are looking at pricing information with a cloud provider online, remember that you are looking at MSRP pricing. There may be other opportunities to further your savings, so we always highly recommend working with your Cloud representatives and/or Microsoft.

Azure Hybrid Benefit is a great way to utilize your already-purchased licenses for Azure and save yourself a lot of money in the process. Remember though that to use this benefit, you must have and maintain software assurance.

Reconsider Architectural Decisions

When you are going through a budgeting exercise, take time to consistently reconsider the architecture. Are you picking the service that best meets the needs of your organization at a cost that makes sense or are there other opportunities worth exploring, such as utilizing different services that would be just as effective at a lower cost? For example, you build an Azure virtual machine to run SQL Server, but you discover you could be using Azure SQL Database for half the price. The calculate, evaluate, and infiltrate method discussed earlier in the chapter is a great way to keep yourself honest here and look at all possible options. Be open to looking at all possibilities.

At some point, you are going to need to make changes to the architecture as a result of cost or functionality. It's always better to handle this as early on as possible to cause the least disruption to the project and to the budget; however, this may not always be applicable. What may be the right solution now might be completely different six months or a year down the road due to the rapid pace cloud technologies change. Consistently reevaluating your architecture can allow you to take advantage of these evolving changes.

Now to be clear, we are not suggesting you uproot your designs to take advantage of something that may save you a few dollars, but if it's significant enough, you might want to consider it. Ultimately, it's your decision to determine if the effort is worth the savings, but at a minimum, you should at least be evaluating these.

Scheduled Shutdown and Startup

One of the most common methods to reduce costs is also one of the most effective. Resources dedicated to lower-end environments typically do not need to be online when not in use. Azure provides tools to perform manual or scheduled shutdown and startup of these resources when not in use. As we've discussed, Azure charges for computer resources by the hour. Identifying and shutting down resources that are not required to run 24×7 is one of the primary ways in which organizations are able to reduce their overall spend.

As with most cost-saving measures, implementing automated shutdown and startup requires careful planning and monitoring. A thorough inventory of your Azure environment and an understanding of what VMs may be shut down are, of course, critical. Effective naming and tagging conventions can assist greatly with this effort (more on tagging is covered in Chapter 2). A good candidate for automated shutdown is any machines in lower environments. Dev/Test/QA resources are often idle overnight and on weekends. While these resources are often priced lower than production VMs, the savings can be substantial.

One important caveat that is also included in Chapter 2 is that it is important to know the difference between **Stopped** and **Stopped (Deallocated)**. An Azure virtual machine in **Stopped** state will still result in charges for compute resources, while one in **Stopped (Deallocated)** will not. We have worked with more than one customer who thought they were saving money only to find out that they had only stopped their VMs, not stopped and deallocated the VMs.

Autoscaling: Provision Only What You Need

The ability to automatically size resources up and down (or out and in) is one of the most profound changes for organizations moving from data centers to Microsoft Azure. In the cloud, you pay for what you provision, not what you use. Hardware is traditionally purchased with substantial unused capacity to accommodate for expected growth. Moreover, peak workloads may occur infrequently but must be accounted for in the new hardware.

Some studies show that average utilization for on-premises resources is around 20%,[1] meaning that companies are paying for substantial amounts of compute and storage that are largely unused. Microsoft Azure provides organizations the ability to provision only what they need to support their workloads and to dynamically scale vertically or horizontally to support growth and peak workloads.

I once worked with an organization in the financial technology business. This organization used transactional replication to replicate over 90% of their entire production dataset to SQL Servers used as data sources for monthly reporting workloads. The entire dataset being replicated was over 1.5 TB in total. Aside from the inherent complexity in running transaction replication, this represented an enormous amount of compute and storage hardware sitting idle for roughly 95% of every month in order to keep the reporting workload from the production transactional systems in sync with the reporting system.

---

[1] https://aws.amazon.com/blogs/aws/cloud-computing-server-utilization-the-environment/

By moving the platform to Microsoft Azure virtual machines and using scale sets, they were able to add virtual machines to support the additional workload and deprovision them when the reporting period was over. The cost savings were substantial, and the customer was able to repurpose the excess hardware to support production workloads.

Another use case can be found in a recent Department of Defense (DoD) hardware contract. The contract specifically stated that hardware would provide at least 60% capacity for anticipated growth of roughly 10% per year for five years. New hardware would be purchased when the utilization threshold reached 75% to allow for spikes and a lengthy procurement and provisioning time. Allowing for the provisioning time, that means that upwards of 20% would *never* be utilized, while much more than that would sit idle for years. For a procurement contract measured in hundreds of millions of dollars, the ability to add capacity as it is needed would result in millions in savings.

Utilize Dev/Test Pricing

Dev/Test pricing is a program applied on the subscription level for Visual Studio subscribers that offers discounted rates to support development and testing efforts. This is a program that saves you a lot by utilizing a program you may already be paying for.

Here are some of the discounts that this program currently offers:

- Windows virtual machines are billed at the CentOS/Ubuntu Linux virtual machine rates.

- Enterprise and Standard Biztalk virtual machines are billed at the CentOS/Ubuntu Linux virtual machine rates.

- No Microsoft licensing charges including Windows, SQL Server, and Visual Studio.

- Logic Apps Enterprise connection is billed at 50% of published pricing.

- Discounts on App Services.

- Discounts on Cloud Services instances.

- Discounts on HDInsight instances.

- Azure DevOps access at no additional cost.

- Access to Windows 10 and Windows Virtual Desktop Service.

To take advantage of this pricing program, you must be a Visual Studio subscriber. If you are not, then you will either need to pay the full price or look into getting signed up for this program. The program is a lot more than just discounted Azure pricing; it also provides access to developer tools, software, support, and training for developers. The current cost of an entry subscription is $1,199 per year with a $799 renewal rate and goes up from there.

Note   To take advantage of Dev/Test pricing, you must be a Visual Studio subscriber.

You will also need to have a separate subscription for your Dev/Test environment and enable this option during the creation process to unlock this pricing. If you are using the subscription that is activated as part of your Visual Studio subscription, then this is enabled by default. If you are utilizing a subscription across your team or organization, then this will need to be explicitly defined. It is also important to note that the subscription must be dedicated to dev/test use only and is not subject to Microsoft service-level agreements. If you are caught running production workloads under this pricing, you could be subjected to penalties from Microsoft.

Azure Reservations

Azure Reservations (or reserved instances) are a commit-to-spend program offered by Microsoft that helps you achieve discounts of up to 70%. You do this by reserving resources for a 1- or 3-year period. There is a little bit of misconception when people hear about these because they think it's dedicated resources, and this is not what it is.

For example, if you have a reserved virtual machine, this does not mean that you are running a dedicated host with just your virtual machines; it means you have committed to use that virtual machine for a period of time and Microsoft offers you reserved pricing as an incentive for a long-term contract.

Note   Azure reservations are reserved pricing, not reserved hardware.

Azure Reservations are a program that we anticipate will continue to evolve, so please check out the latest documentation on these before making a decision. However, we do want to share some important insight and consideration in using these.

First, you need to be aware **that Azure Reservations are something you can easily self-manage via the Azure portal**. This is not a program that requires you to interact directly with Microsoft to take advantage of, although it is something that you can absolutely reach out to them for assistance on. Just search for Azure Reservations, and you can start creating and managing these as shown in Figure 3-8. You will want to check that screen or the official documentation to see which services are applicable, since not all services are.

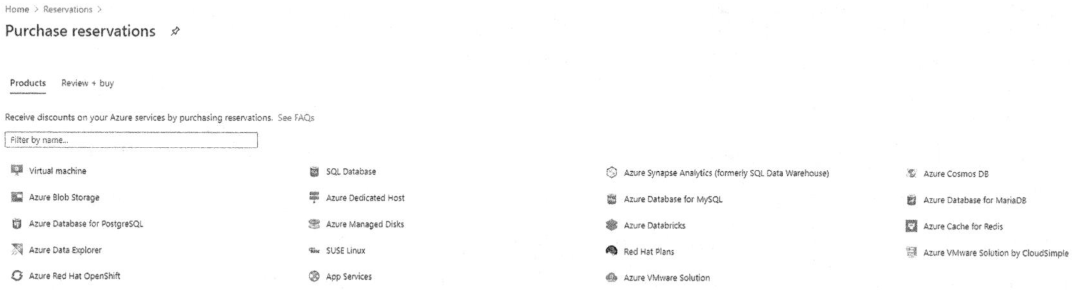

Figure 3-8. *Purchasing Azure Reservations in the portal*

When you create a Reservation, **you will need to define the scope**. A *scope* is the layer within the Azure Entity Hierarchy that you wish to apply the reservation on. Currently, you are limited to setting the scope to one of three options. We will review each of these individually:

- **Shared**: This allows you to share the reservation for matching resources across multiple subscriptions and can vary based on the type of agreement you have. This is usually the default choice because it gives you the widest coverage.

- **Single subscription**: This allows you to apply the reservation to matching resources in a single subscription. You may want to consider doing this if you have multiple subscriptions for different purposes and you want to make sure your credit only applies for your resources.

- **Single resource group**: This allows you to apply the reservation on the matching resources on the resource group layer. This is currently the most granular layer where a reservation can be applied. You may consider doing this if you have resource groups in a subscription being used for different purposes and want to make sure the credit is only applied to a specific area.

If you do create one and determine that you need to widen or narrow the scope of a reservation, then you can go into the portal and just change this at any time.

Reservations are applied on an hourly basis for every applicable resource except Databricks, and the discount is use it or lose it. If you are running 24×7 workloads at a static service tier, then you will be able to get the full benefit. If you are pausing and resuming a service, you will lose the discount applied during the times you are not running. You still save by not having to pay for compute resources during that time. But your overall saving percentage is going to be lower, and you need to factor that into your reservation.

If you are creating scale points to scale the service up or down during times, then it will also have an impact on your discount. Typically speaking, you will only want to reserve resources up to your lowest tier running to avoid overcommit. This does still result in savings, but it's not going to be as high as the discounts advertised.

Note   Azure Reservations work best for predictable 24×7 workloads.

I will share an example of this using Azure Synapse (formerly Azure SQLDW). This is an expensive resource that we are using to host our data warehouse in Azure. We have heavy batch processing that occurs from midnight to 6 a.m., so we run a 2000DWc tier to accommodate this workload. From 6 a.m. to noon, we have several report refreshes and data extracts happening that require a lot less compute, but still some. So we scale down during that time span to a 1000DWc tier. Then from noon to midnight, we just have light querying and reporting that happen, so we further scale down to a 500DWc. In this scenario, we have saved our organization significant dollars by running at the scale needed to handle the workload rather than just setting it to a 2000DWc and leaving it there.

Now consider if you want to apply a reservation to this scenario because you are told that you can save up to 70%. But this presents a challenge because the reservation is applied on an hourly basis. And you are reserving capacity which you cannot apply the discount to the 2000DWc service tier because you are only running at that workload for six hours of the day, not 24. Doing this would result in an overcommit.

Instead, you would apply the reservation to the lowest common denominator of matching resources in your scope. If this is the only Synapse pool that we are running, then the max reservation should be DW500c because it can be applied effectively each hour. This will mean that during the time we are running a higher workload, we are paying full price for anything beyond DWc, significantly lowering the percentage of discount we get.

If we had a scenario where we knew we would at least be running a DW2000c 24×7, then we would set the reservation to that level and be able to maximize our discounts. There is some strategy and calculation that should occur if you want to maximize your benefit. However, it is a worthwhile experience, and it can be very beneficial to do so.

The next consideration is that **you can choose to pay up front or pay them on a monthly basis**. There is no penalty currently if you choose to pay monthly, so it's up to you to decide what works best. Either way, you will get the discount applied.

You will want to **look at the official documentation to understand cancellation fees before committing**. As of this writing, there are no cancellation fees for canceling these, but Microsoft has already hinted at a 12% cancellation fee being added in the near future. They already have a $50,000 cap on refunds, so you will really want to evaluate before making a decision.

You will also want to **look at the official documentation for each service to determine what is covered as not everything is**. For example, storage and software licenses are a notable exception for several services that discounts are not applied to, although compute generally is included, and this oftentimes is the most expensive cost anyway. It will still be important to understand what your discounts are applied to versus what they aren't.

To be efficient with Reservations, **we recommend waiting to apply these until your workload has stabilized and you know it is not going to change**. During the budgeting phase, you can evaluate the pricing for these using the Pricing Calculator, but it usually isn't a good time to factor these because things can change. Instead, build out your workload and give it enough time to stabilize before considering these. You will want to make sure you can live up to your commitment first before going through with it.

The Non-Azure Factors

The budget for an Azure migration is not exclusive to just Azure resources, and there are other items that also should be considered and factored into the budget.

The first of these is any **additional software licenses**. These could be things like monitoring tools or security software but could be any software that requires a paid license that is not already part of Azure. These are primarily going to apply for virtual machines that you are running in the cloud, because you retain responsibility for anything on the OS for them. This includes anything that needs to be installed on them and any subsequent licensing or fees needed.

In some cases, you may need to procure a special license to run in the cloud or may be able to transfer a current one. In other cases, you may find the software you are planning to use is not supported in the cloud and you need to utilize something different to retain the functionality. You want to be sure to factor any costs for those different scenarios.

For some options, Azure may have functionality that you can use, but maybe not as robust as you would like and you want to pick something else. Monitoring tools are a good example of this. The cloud does have some baked-in monitoring functionality, but you may still want to include something else that provides you greater details. For example, you may still want to monitor your database workloads using your favorite third-party product to gain deeper insights and greater functionality. And you can absolutely do this; you will just want to factor in what that cost means to your project. This topic is covered in greater detail in Chapter 5 if you want to learn more.

The next item is **training for your team**. Migrating to the cloud is a big undertaking that requires a lot of skill to be able to use. A common mistake that we have seen is that companies spend thousands or even millions of dollars to move to the cloud and don't invest in making sure their staff can support it. This is not only poor management, but something that can lead to costly mistakes. When building out your budget, you will want to factor in dollars to be able to get your team up to speed. This is another topic that we go into a lot greater detail later in the book.

Building an ROI

In this section, we want to highlight some of the items that you can use to sell the return on investment (ROI) to your organization. These are a list of selling points that can be used to communicate to your decision-makers why this project is a good idea.

The first of these is **being able to determine the costs of moving to Azure and comparing that to your on-premises costs**. This is essentially using the information provided in this chapter to create a budget and then doing some comparisons of how that compares to your current costs. You can use the TCO calculator to provide a deeper comparison by looking at all factors or you can first focus on the big-ticket items to see if that is enough.

The next item is being able to **shift Capex dollars to Opex**. *Capex* stands for capital expenditure and is expenses that are paid now to generate value in the future. This is going to include things like hardware and equipment purchases. *Opex* stands for operational expenditure and is expenses related to the day-to-day operations. This is going to include things like services or cloud spends. Generally speaking, organizations prefer Opex to Capex because of the ways they are accounted for on an income statement. A Capex expense cannot be fully deducted in the year they are incurred because they are amortized or depreciated over the life of the asset. For something like hardware, this could spread the investment out over a three- to five-year period. An Opex expense is fully deductible within the same year. This is important because most companies are taxed on the profit they make so they want to deduct as many expenses as quickly as they can to lower their tax bill.

For organizations, having more money now is more important than later due to its earning capacity. There are scenarios where Capex will be preferred, and this usually occurs when organizations are trying to produce a higher value of assets on the balance sheet, which results in a higher net income that can be reported. If it is important for the organization to shift more costs to Opex, then this can be a big selling point. However, if a company prefers Capex spend, then reservations can be a big selling point as well.

The item is **utilization and scale**. If you are procuring hardware, you are doing it for years at a time, so you have to project what you think you need to be able to support the business. This can often lead to a case of over- or underprovisioning. If you overprovision, then you have wasted dollars for something you don't need. If you underprovision, then you risk constraining the business while you procure and install additional hardware which can take months. And even if you get it right, think of all the hardware that you have running in the event of failures.

It would be common to see hosts in VM farms running at fractional capacity, because you have to be able to handle the workload in the event of hardware failures. Within the cloud, you only have to pay for what you need when you need it. And if you make a mistake, it can be remediated within a few clicks as opposed to wasted dollars on unnecessary hardware or months to get what is needed to expand.

Note   Within the cloud, you only have to pay for the resources you need when you need them.

Time reduction can be another key factor for several of the reasons we discussed with utilization. This is because we can create, modify, and drop resources in the cloud significantly faster than we ever can on-premises. The time to deployment in Azure can occur in as little as seconds. Let's say that your organization signed a contract to take on a new client that would double your existing volume. You are probably going to want to get them onboarded as soon as possible because it amounts to a lot of increased revenue. The cloud makes this possible.

Let's say that your organization is in the business of purchasing assets. Wouldn't you want to have the ability to quickly onboard them without delays related to infrastructure? Again, the cloud makes this possible. The reduction in time of being able to do things significantly faster than ever before can be a huge selling point to your organization.

The next thing is having **access to cutting-edge technology without needing to do anything to get it**. In the world of today, executives and teams are getting wowed at conferences and briefings with all the cool things we can do with data. Things like Big Data, Analytics, and AI are sweeping throughout industries and changing the way companies operate. These things look cool but can vary in their approach and utilize very different technologies. The cloud provides us a place where we can begin exploring these, as well as other cutting technologies without the cost of a major purchase.

There are several other cloud benefits that you can use to sell the idea to your organization such as **security**, **accessibility**, **simplified administration**, **improved high availability** and **disaster recovery**, and more. You can take any benefit associated with the cloud and use that to showcase a return on investment for your organization. These items will be detailed further in other portions of the book.

Summary

This was a long chapter that introduced us to the concept of cost management in Azure and provided valuable information on how to build a budget for your Azure migration. We reviewed how essential the Azure Pricing Calculator is to the budget building process and the many important considerations both before and after you needed to account for. We also looked at existing programs and expert tips you can further optimize the costs to get even better cost savings. Building a budget will require some time and effort, but it is not an impossible task. With good planning and diligence, you can be better prepared to answer important cost questions and get your organization on board with the next steps in your cloud migration.

CHAPTER 4

Azure Cost Management

If cost management is important to you, then it's not enough to just estimate the costs; you also need to be able to manage your spend throughout your cloud journey. Because the cloud is so dynamic in its pricing, you are going to have to manage things to be sure you can stay within your budgets. Or understand as soon as possible when you may have missed the mark. In the previous chapter, we looked at the first part of cost management, which is building a budget. In this chapter, we are going to look at the second part, which is how to manage your costs once you are already migrated.

Azure cost management is not only just a concept, but it's also a tool that is part of your Azure tenant that can assist in your goals for good financial governance. In the Azure portal, this is referred to as Azure Cost Management + Billing and combines together a suite of tools to assist you with managing your billing and effectively managing your costs. These tools, which are free for all Azure customers, allow you to monitor your cloud spend through interactive data and reporting, generate alerts and automated actions, and assist you in optimizing your cloud costs. In this chapter, we will review important cost management concepts, but we are primarily going to focus on the features of Azure Cost Management tool within the Azure portal and look at how it addresses key components of cost management.

Azure Cost Management and Billing

Before we dive into the details, we want to introduce you to Azure Cost Management + Billing in the Azure portal by looking at how it evolved and what its intended use is. Back in 2017, Microsoft acquired Cloudyn, a company that helped customers manage their billing across multiple clouds using a cost management application. Microsoft started taking the cloud billing and management application Cloudyn had already produced and using it to build on its functionality within Azure Cost Management. Much of the primary functionality of Cloudyn has already been transferred over with some lesser

© Kevin Kline, Denis McDowell, Dustin Dorsey, Matt Gordon 2022
K. Kline et al., *Pro Database Migration to Azure*, https://doi.org/10.1007/978-1-4842-8230-4_4

components being slated for later releases. Microsoft has already stated that they will not be onboarding any new customers on Cloudyn and that it is currently slated to be deprecated by the end of 2020. You may still see a link to the page in the portal even, but it's not something you want to use. It was still part of it because of the customers that still needed to be migrated off.

Azure Cost Management + Billing is not an old product that we have had since the early days of Azure and really only began to take shape after the acquisition. Before that, you had access to invoices you could view, but this was about the extent of it. And being able to optimize using an itemized invoice with Azure's pricing model is a tall order for anyone, which is why third-party tools were often used. Thankfully, Microsoft invested in this easy-to-use solution that gives us so much deeper insights into our costs. There is still a lot of room for growth with the toolset, and we expect to continue to see more added to the product over the coming years.

You should also be familiar at this point that Azure uses a pricing model that is "pay for what you use." This means that you are only paying for the resources allocated or consumed. Each time that you create or alter a resource, it affects your spend and ultimately what shows up on your invoice. This sounds great for anyone new to cloud because it immediately sounds like you are going to save money. However, it can be a double-edged sword that can bite you without proper oversight. This suite of tools is designed to help avoid that and make sure you are getting the most out of your cloud spend.

Azure Cost Management does have several administrative functionalities, but it's also designed to help you be proactive with managing your costs. Here are some of the important features that Azure Cost Management + Billing offers:

- Conducting billing administrative tasks such as reviewing and paying bills

- Managing billing access to costs

- Exploring organizational cost and usage patterns

- Ability to analyze costs using interactive reports combined with advanced analytics

- Creating budgets and alerts with automated responses

- Predictive analytics

- Analyzing the impact of architectural and technical decisions

Azure Cost Management also consists of four primary components of functionalities: billing management, analysis, budgets, and advisory recommendations. Billing management is where you can view and pay your bill. Cost analysis is where you can really dive in and analyze your data through interactive reports. Budgets are where you can drive accountability by setting thresholds on your spend and automating actions when percentages of those are met. And lastly, advisory is like your personal cloud consultant that provides recommendations on ways you can save money. We will go through each of these in greater detail throughout the chapter.

Whenever you begin using Azure, you must do so under an existing or new customer agreement. A Microsoft customer agreement is simply a purchase agreement, and there are several different types to fit the needs of differing organizations. Within Azure Cost Management, the most common ones are supported; however, some of the others may not be yet. Let's look at the agreements that are currently supported.

- **Microsoft Online Services Program**: This is an individual billing account and is what is created if you create a new account directly via the portal. Typically, this is going to be a pay-as-you-go account.

- **Enterprise Agreement**: A billing account for organizations that have a signed Enterprise Agreement with Microsoft.

- **Microsoft Customer Agreement**: A billing account that is used when you work with a Microsoft representative to sign an agreement.

- **Cloud Services Provider (CSP)**: A billing account that is used when you have a partner managing your entire Microsoft cloud customer life cycle.

If you are under a different agreement with Microsoft, we would recommend checking the official documentation to see if it's supported now or reaching out to your Microsoft representative to discuss with them. It could even make sense to change your agreement type to something that fits your needs better.

It is also important to note that based on your customer agreement, it may slightly change some of the experience in Azure Cost Management. For example, you may have additional financial layers that you can roll your data up under or additional financial metrics to analyze your data that are specific to your type of agreement.

Cost Management Concepts

There are four pillars that make up the concept of cost management that you need to be aware of. These pillars are illustrated in Figure 4-1, and they are planning, visibility, accountability, and optimization. They are each ongoing activities that you should do throughout your cloud journey.

While the activities in Figure 4-1 each represent a way of thinking, they also translate to real functionality within the tools that we will be covering throughout the chapter. It's important that this becomes part of your methodology as early as possible and ideally before you spend any money on any cloud resources. However, if you have already started, it's okay and never too late to adopt.

Figure 4-1. *The Cost Management Concept life cycle*

It is also important that you are aligned throughout your organization on what this process entails. This is not a one-time process, but an iterative process that needs to become part of the team's strategy going forward. Whether you are a member of finance, a business decision-maker, or a member of a technical team, it's important for everyone to work together to create good cost governance.

Let's look through each of these concepts individually in more detail.

Planning

One of my favorite things to do is travel with the family, especially going on vacations. We always try to do at least one big fun family trip per year. Each year, I put some serious time into planning these trips because I want it to be the best that it can be. And also admittedly, because I love the planning aspect of it. Having young kids, a place we have visited several times is Disney World in Florida. For this trip, it is a must to start planning very early, and we often start nearly a year in advance. Otherwise, you could miss out on some important aspects of what makes your vacation magical such as finding open occupancy at your favorite hotel or getting those hard-to-get dinner reservations or booking fastpasses for the most popular rides. Sure, you could just decide to go on a whim with no planning and still be able to get into the parks, but at what cost? Skipping the newest ride because wait times are astronomical? Settling for burgers from a quick service because the magical princess fairy dinner your kids want so bad is booked up? I know there are people who frequently visit, but for us, a special event when we go, so getting the most is imperative. Migrating to the cloud is a lot like taking that special trip to Disney World. Sure, you can just go without planning, but at what cost? Spending the time to plan things out can significantly improve your experience and save you a lot of headache. You don't want to be in the middle of your migration having regrets because you didn't properly plan in the beginning.

Good planning is one of the most important concepts to cost management because it sets the stage for everything else you do. Failure to do so can make things more difficult when you start. Good planning in relation to cost management consists of two essential components: *estimating your costs* and *creating good cloud governance*. Estimating your costs is figuring out what you are going to be migrating and what the estimated costs are associated with that. Without having a starting point, it's going to be very difficult to tell how you are tracking against expectations. We covered this concept in great detail in the previous chapter. If you have not read through that, we would encourage you to do so. Going through the process laid out in the budgeting chapter is going to be essential to success with cost management in the portal.

Note   Good planning with cost management consists of estimating your costs before starting your migration and creating good cloud governance.

The next component is creating cloud governance, which is about building a foundation that supports you throughout your cloud journey. What is governance? Governance refers to the process of managing, monitoring, securing, and auditing your cloud environment to meet the needs of your organization. This is really about the foundational elements that make up the framework you build your cloud infrastructure on. It is a designed set of rules and protocols put in place to enhance data security, manage risks, and keep things running smoothly.

Having a good governance strategy is essential to your cloud migration and one of the most important things you need to do to have a successful migration. This is a big topic and something we dive into greater detail with elsewhere in the book. However, since it does play a crucial role in cost management, we will review it from a high level here. Cloud governance can be broken down into four key areas that we will walk through, as well as their impacts on managing costs. The four areas are

- Resource organization

- Resource security

- Auditing

- Cost controls

Resource organization covers a wide range of topics and can impact your Azure Cost Management in the portal experience. First, it is how you structure or containerize the allocation of resources in things such as subscriptions, resource groups, and management groups. It also includes the naming convention you use for your resources. These are really important for cost management because they are components in how you navigate through your cost data. If there is no structure or thoughts given to how these are implemented, then you are going to limit yourself on being able to analyze your costs.

A specific item related to resource organization that we want to mention is tagging. *Tags* are user-defined metadata that can be added to most Azure resources. These allow you to be able to organize resources based on your own criteria and can really enhance the cost management experience. Tags can be things like resource owners, departments, environment details, application, expiration dates, or any other type of data you want to view your resources by. Using descriptive naming conventions is great practice, but you are still limited there. This opens a whole new area of possibilities for organizing your resources in ways that make sense to your organization. You may still use tags overlapping things that are part of your naming convention. You can also create policies to enforce tagging, so anytime a resource is created, it requires values.

Note   Tags allow you to be able to organize resources based on your own criteria and can really enhance the cost management experience.

The next phase of good governance is resource security. This refers to things like role-based access control (RBAC), Azure policy, and resource locks. This is important to cost management because one of the best things you can do is to limit who has access to make changes that can impact cost. If everyone on your team has elevated rights, then anyone on your team can make a decision that could impact the underlying cost and puts you in a greater position of someone making a mistake. The beauty of cloud technologies is the ability to do things quickly and enable your teams; however, you still need guardrails. There are a lot of approaches to how to manage cloud access, and it's ultimately up to you and your organization to decide how to implement it.

Auditing is a big topic and consists of several aspects that are discussed in other areas of the book. This is essentially about having methods in place to track and evaluate activities in your cloud environment, as well as adhering to internal and federal security protocols. In terms of cost management, audits can often answer the question of why and can assist with creating accountability throughout your organization.

And the last component of good governance is cost controls, which is being able to manage your costs. This is what we are discussing throughout this chapter so a lot more to come on this.

If you are unsure where to get started with your Governance strategy, Microsoft has several resources available that can help you. First is the Microsoft Cloud Adoption Framework, which is available for free online (`https://docs.microsoft.com/azure/cloud-adoption-framework`). Another is via the Quickstart Center in the Azure Portal. Simply log in to the portal and search for a quickstart center if you do not already see it and go to the Azure setup guide. It provides documentation and step-by-step instructions on getting started.

Visibility

Creating visibility is having insights into your cost data and using that information to make everyone aware of the financial impacts of their solutions. This starts with having visibility into what the financial impacts are. If we don't have tools or data that provides that insight, then no one can be made aware. Using tools like Azure Cost Management provide data to help keep people informed about the impact to cost they are having. If a person has no insight into the cost impact of decisions they make, then the trend of any bad decisions can never be rectified. Visibility is all about making people aware of the impact of their choices.

Azure Cost Management in the portal provides insights into showing you where your money is being spent through advanced analytic reporting. Here, you can navigate through your data to begin to answer the questions of who and how to educate. Using these tools, you can dive into the details of the cost to understand areas that could be underutilized, wasteful, or non-optimal.

Accountability

Whereas visibility is about making people aware of their impacts to costs, accountability is about holding people responsible and working to stop bad spending patterns. You need to be able to determine who or which team is responsible for those patterns so that they can be addressed. We are not talking about walking around like a drill sergeant calling people out and getting them in trouble. The cloud is a new technology and way of thinking for a lot of people, so it's going to take some time to adapt. The goal with visibility is helping them understand the impact of their actions so they can learn and not repeat them.

Discovering who made an impactful change starts with having good resource organization governance, structuring your resource organization and security in a way that supports your organization setup and limits and spotlights the responsible team. An example of this would be having subscriptions or resource groups limited to specific teams and using naming conventions that clearly indicate that.

Within Azure Cost Management in the portal, we identify the issues through the analytic reporting or we can automate these checks using budgets. We will look at each of these in greater detail later in the chapter.

Optimizations

Optimization is about finding ways to do more with less and looking for those cost-saving opportunities on a consistent basis to lower the costs. This includes evaluating whether one of the Microsoft cost-saving programs makes sense to utilize, following tips laid out throughout this book, or evaluating Advisory recommendations. This includes such things as finding a service that was inadvertently left running 24×7 or incorrectly provisioned.

Optimization is on the forefront of everyone's mind when utilizing Azure Cost Management, and it is ultimately why they are using it. No one who cares about cost management wants to spend more than they need to. As we dive more into the tool later in this chapter, we will begin to see a lot of ways we can discover optimization opportunities.

Azure Billing Entity Hierarchy

When you first access Azure Cost Management, you will need to select your scope. A *scope* is a node in the Azure resource hierarchy where Azure AD users access and manage services. Each item that you can select for your scope represents a place in the Azure Billing Entity Hierarchy. The way the hierarchy is structured is by billing entities and resource entities. The billing entities are based on your Microsoft agreement and will vary. The resource entities are based on your resource organization and will be the same across all agreements.

In the hierarchy, the resource entities always roll up underneath the billing entities. The billing entities would consist of the billing account and any additional billing items related to your type of agreement. Here is a list of the specific entities based on the agreement types:

- **Individual Agreement:** Billing Account

- **Enterprise Agreement:** Billing Account, Departments, Enrollment Account

- **Customer Agreement:** Billing Account, Billing Profile, Invoice Section, Customer

- **Cloud Solution Provider (CSP):** Billing Account, Billing Profile, Customer

The Billing Account for each of these is the root for all Billing Entities and exists in any agreement type. Each of these entities represents a place to view your billing information in Azure Cost Management by changing the scope to this.

Resource entities would include items related to your resources or the different layer of items that make up your resource organization governance strategy. At the top of the resource entities would be any management groups if you are using them. Management groups are containers that assist you in managing multiple subscriptions and are often created to assist in managing access, policy, and compliance and grouping together data for cost analysis. Management groups are another layer of strategic containerization. A simple example of using this is if you have your production and development subscriptions separated. A reason you may decide to do this is to take advantage of Dev/Test pricing on your development resources to lower costs. But you may still want to be able to manage access and policies across both and see costs collectively across both subscriptions. Creating a management group that contains both the development and production subscription allows you to do this. This is a simple example, but there would be several other instances where this would make sense.

Underneath the management groups, you have your subscriptions, followed by resource groups and then resources. The higher you are on the hierarchy, the more widespread your view, and the deeper you go down the hierarchy, the greater the level of granularity you achieve. Your scope represents the layer at which you want to view your data and can be changed at any time to represent your desired layer of granularity.

Azure Cost Management in the Portal

We have introduced the concept of cost management, as well as Azure Cost Management + Billing in the portal. We are now ready to start looking at the actual functionality within the portal to see what all you can do via the tools. We will be looking at the areas of Cost Analysis, Budgets, and the Azure Advisor.

To access Azure Cost Management, open the Azure portal and locate the search bar. In the search bar, type "cost management" and you should see the service. Just click on it to access Azure Cost Management. You can access it in other ways as well, but this is the easiest way if you are doing it for the first time. Figure 4-2 shows the Cost Management home page once you connect.

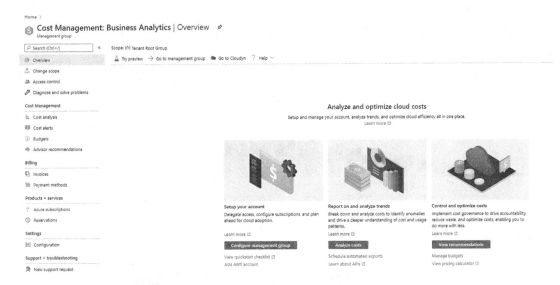

Figure 4-2. *Azure Cost Management + Billing home page*

Cost Analysis

Cost analysis is the part of Azure Cost Management that allows you to navigate through your costs using advanced analytics via an interactive dashboard shown in Figure 4-3. When you first access this, you will see the main chart at the center of the page with several smaller pie charts listed below it that displays your cost data. From here, you can dive into your data and begin to analyze it to evaluate your current spends.

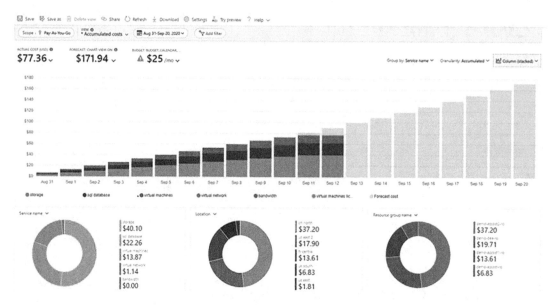

Figure 4-3. *View of Cost Analysis in Azure Cost Management + Billing*

When you first access cost analysis, you will first need to select your scope. It will usually default to the highest level in the Azure Billing Entity Hierarchy that your account can access, but you may need to adjust based on the level of granularity you are interested in. Currently, the scope is the top left-hand page of the screen as demonstrated in Figure 4-3. Just click on it to set it at the appropriate level. Earlier in the chapter, we discussed the Billing Entity Framework, and this is where you can navigate that. Once you set your scope, you will see the data on the page automatically get updated to reflect the new scope.

From here, you have several options on how to start slicing and dicing your cost data, as follows.

Change the date based on several different built-in date ranges or use a custom date range. This includes things like year to date, this month, previous month, previous week, and so on.

Add filters or change the grouping using any of the resource or financial dimension attributes available. The attributes will include items that make up the Billing Entity Hierarchy and also include several other specific items related to the resources themselves. This will include things like service names, locations, resource types, and more. And if you are filtering data by a specific attribute, you don't have to type in the value you want; rather, everything is selected via drop-downs. Also, while you can only add one grouping per report, you can add multiple filters.

Because you are filtering and changing groupings of your data with this, you can see why having a good resource organization governance strategy is important to this. Having a standard structure to these with a naming convention that is consistent and makes sense can make evaluating your costs a lot more fruitful.

In addition, earlier in the chapter, we looked at tagging. With filters and grouping, you can use tags attached to your resources to be able to analyze data beyond the native capability. This opens up a world of new possibilities to begin exploring since you can create tags to support any data point you like. These can commonly be things such as environment types (prod, dev, etc.) or information about who owns or created it. Another commonly used tag is departments or business owners since these can be used to perform chargebacks. Chargebacks are something that some companies do to allocate spend to specific budgets. If you have multiple departments in a tenant working under the guidelines of their budget, then you need to be able to distribute the costs accordingly. This is a common use case that companies deal with, so expect to see improvements made to this area in future releases.

Change the granularity of the reports. This determines the granularity at which you want to view your cost data. Currently, you change it to accumulated, daily, or monthly. Accumulated shows your ongoing costs, so each day would include the costs from the previous days, as well as any new charges. Daily would change the costs so that you are viewing what your actual spend was for a specific day. The Monthly view would change the costs that you are viewing to show your total costs on a monthly basis.

Change the chart type on the primary report. The report will default to an area report, but you do have the option to change it to other types. The current options are area, line, column (stacked), column (grouped), and table. The table view is text data in a table format. You would use this if you want to see data in a more raw format.

The three donut charts that are located at the bottom of cost analysis cannot be changed to another report type; however, they do reflect any changes that are made to filters or grouping. You also can select what grouping is used for them that differs from the primary report on the screen.

Change the view of costs to an amortized view. By default, the cost data is shown as actual cost. However, if you are using Azure Reservations, you may want to view it differently. Because reservations are applied on day one of the term, you would see all of it at once on the actual cost view. Changing it to an amortized view breaks down the purchase charges and spreads them out evenly over the lifetime of the term.

Let's look at an example. Let's say you purchase a 1-year reservation for $12,000 on January 15 and choose to pay for it monthly. You could choose to pay it all up front, but let's go with the monthly option since it will have a greater impact on how you view costs. Starting on January 15 (the date of purchase), you are going to see a charge of $1,000 in cost analysis. And each month following on the 15th day of the month for 1 year, you are going to see the same $1,000 charge. Because you are paying for it monthly, you may not want to display it that way and want to balance that charge over the course of the month. This is what the amortized view does. It takes $1,000 for the month and spreads it out evenly for each day in it. So instead of seeing a lump charge for the month, you would see daily charges for it.

Depending on your overall spend and the size of your reservation, the actual cost could really inhibit your analysis so you have a way to change should you like.

Add filters by clicking directly on the reports. The reports themselves are also interactive, so you can click directly on them to filter the data. For example, if you have a donut chart at the bottom of the screen that shows locations (or regions). You can simply click on the region you want to only view data for, and a filter is automatically added, and all reports are updated to only reflect data for that single location.

Being able to forecast spends. When you are navigating your data and have future dates in your view, then you will automatically see forecasted spend. It uses the consumption data based on the previous dates collected to project out your spend. Let's look at a simple example. The month of November has 30 days in it, so let's suppose on November 10 we had a spend of $10,000. We then went into cost analysis and looked at our cost data for the month of November. The forecaster would evaluate our consumption and estimate that based on current trends, we would spend $30,000 for the month. We could make a determination to see if this was in line with our budgets or if it was going to be a problem. Because it's still early enough, we would have time to remediate this if we needed to. In the real world, costs may fluctuate and may not be as simple to figure out, so the forecasting feature helps with this. Additionally, as you dive through your data, the forecasting changes based on your selections, so this creates some very granular insights even down to the specific resource layer.

Viewing spends against budgets. We have not looked at budgets yet but wanted to go ahead and mention it here. We will review this within the next few sections.

Ability to share and export data. There are several ways to get the data out of cost management if you want to be able to save them or share with other members of your organization. You have the ability to download images of any of the data or charts that you generate in a PNG, Excel, or CSV format. You can also generate a link for any reports that you have created that can be shared and viewed by anyone who has access. You can also schedule exports of your billing data to a storage account on a daily, weekly, or monthly basis for historical tracking or to share out. As of this writing, the ability to set up email exports and automated delivery is not supported; however, you can still set this up if it's important to you via other methods.

Cost analysis is the foundation of Azure Cost Management and the place that you can analyze and evaluate your cost data.

Getting the Most from Your Analysis

Having the ability to analyze and evaluate your costs is useless unless you understand how to get the most from it. When we review our organization's cost data through cost analysis, these are things that we are constantly looking for and questions we ask ourselves about it. These are things you can use to help you throughout the process.

The first thing that we look at is the **actual costs versus our budgeted costs**. Here, we want to see how we are trending based on our budgeted amounts to our actual spend. You should have an idea of the expected costs of your project based on the work done to build your budgets as a point of comparison. If you don't have a budget, you need to create one even if it's later in the project so you have an idea of what normal spend should look like.

Look to see how you are trending and be prepared to do further analysis if things look to be going off the rails to understand why. You always want to get ahead of any potential overspends well before our bill comes so you can adapt or begin preparing our leadership team on the increased spend. No one wants to have a surprise when you get the bill.

The next thing that we **look for is any anomalies**. Anomalies would be sudden unexpected spikes or drops in costs accrued. A spike could mean that someone unintentionally left a service running or scaled a resource up unbeknownst to other members of the team. A sudden drop could mean a service was deprovisioned or left off. Minimal changes are expected, but if we see a significant difference, then we need to investigate that. The idea here is to look for anything unusual that could evolve into a bigger issue later.

The next thing is **reviewing invoice comparisons**. This is looking at how your spending this month compared to the previous months and seeing if any of your spending habits have changed. It is normal that cloud costs increase over time if you are evolving and building more and more, but you should be able to explain what those are.

The last thing we will typically look at is **how chargebacks are distributed**. It is common to have multiple groups with different budgets working under the same Azure tenant. Each group is responsible for staying under their allocated budget, so it's important that you are not paying for something someone else should be. When finance gets the invoice from Microsoft and requests approval on charges to your department, you want to be able to verify those. We understand that different organizations handle budget allocation very differently, and for some, it is more important than others, so I'll leave it to you whether this is something you check or not. As mentioned earlier, an easy way to do this is via tagging and/or creating a resource organization that supports this.

Proactive Performance Tuning

All of the writers of this book have spent a considerable portion of their careers as database administrators (DBA), and this is how we started our careers in working with data. One of our favorite and most rewarding aspects of being a DBA/data professional was performance tuning. And even in our current roles, this is still something we still get to do and love even if we aren't as dedicated to it as in the past.

The companies that we have worked for throughout our career have consisted of small DBA teams, so a lot of the work we did was project based or very reactive. Because we often had so much on our plates already, being able to proactively performance tune was not something we got to do very often. Certainly, we did a lot of performance tuning, but it was often because of a fire that needed to be put out as opposed to just improving a process. We can recall several conversations throughout our careers with managers and leadership about staffing to be able to do this and the potential cost savings associated with doing so, and it usually did not lead to much change. Why? Because most of the cost savings related to performance tuning for them came from saving on licensing, which were costs already consumed. The savings were more of a cost avoidance over a cost savings, which caused it to be viewed very differently. Sure you can save on maintenance costs for software assurance, but this cost was usually negligible compared to the overall spend. Ultimately, it was never something viewed as priority. We also understand this is not the experience of every DBA in every organization out there and there are industries that require proactive performance tuning, but there are several that do not and we spent time with some of them.

If you are part of one of these types of companies, then be prepared because the cloud will likely change this way of thinking. Because in the cloud, everything comes with a cost that isn't consumed in an up-front capital expenditure, meaning that you are paying for these services as you are using them. This means that if we can have our team consistently looking for optimization opportunities on something that is one of our most expensive costs, then why wouldn't we?

Let's assume that the average life span of a database in your organization is five years (which is probably really low) and you have a database that at the beginning of its life you want to move to the cloud. So you move the database for the application to a 16-core General Purpose Azure SQL Database running 24×7, which is costing you around $2,000 per month. If we can tune the queries in that database to drop it from 16 cores to 8 cores, then we have just cut our monthly spend in half for this resource. That is, a $1,000 per month savings that spread over the course of five years is $60,000. And that is one database! If you can do that across the board, what would potential savings look like? We believe there is a big difference from the perspective of leadership on cost avoidance versus immediate cost savings. And some of the mentality of not spending time on proactive performance tuning is changing for a lot of cloud adopters. Speaking from experience, we know this is something that we have seen viewed very differently.

So you might be thinking, well why are you telling us this here in this portion of the book? Well, the answer is because cost analysis is a way to validate the difference you are making and promote more proactive performance tuning. If you are able to tune a resource and drop the cost, then you now have real data to go back to that translates to your monthly invoice that you can show to your leadership. You can share more irrefutable evidence of the value of time you spent working on something and to us that is a beautiful thing.

Note   Cost analysis is a tool that can be used to show the financial impacts from your performance-tuning efforts.

Additionally, if you are a database administrator that is afraid of cloud migrations due to job security, then think again. There are plenty of skills that translate to the cloud that are essential, and performance tuning is one of those. We believe this is a skill that will now be more important than ever and could even be the deciding factor on whether a migration is successful or not.

Budgets

So you might be looking at the previous sections on cost analysis and thinking you don't have the manpower to be watching the charts and analyzing data constantly. Well, you wouldn't be alone. This is where budgets play a vital role. *Budgets* are configurable thresholds that can be defined in Azure Cost Management that can trigger alerts with automated responses when percentages of those are met. Budgets work like an extension of your team without needing to hire an additional person or requiring someone to be watching it constantly.

Note   Budgets can be an extension of your team without the need to hire additional resources.

To access these, you just navigate to Azure Cost Management and select the option for Budgets. You'll see a page like that in Figure 4-4. There are no budgets created by default, and it is up to you on how you want to create these.

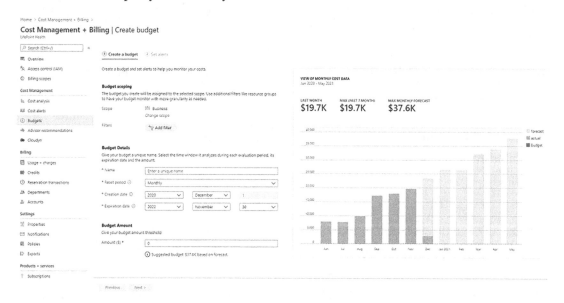

Figure 4-4. *Example of creating a budget in Azure Cost Management*

The first thing you will need to do when creating a budget is define the scope. Again, this is the layer of the Billing Entity Hierarchy or granularity you want to create the budget on. Next, you will need to provide the budget details. First, give it a unique name that describes what the budget is being used for. Next, you will define the reset period. This is the time windows analyzed by the budget and can be monthly, quarterly, or annually. The next parts are the dates in which you want the budget to be active. This includes a start date and an expiration date. You can start budgets in the future, so if you are budgeting for the next year, you can add that in ahead of time. The last step is setting the dollar amount for your budget.

When you set the dollar amount for your budget, it will provide you with a suggested amount. This is a forecasted spend amount based on what your previous spend has looked like, so do not just default to this. You need to evaluate ahead of time the resources you plan to deploy and determine the estimated cost. There is also a chart view that is shown that will add a line to show you how your budgeted spend correlates to your forecasted spend. This is a helpful visual, but again, we would not recommend building your budgets based off of this.

We recommend always starting with adding your monthly budget early so you can stay ahead of your total spend and then creating additional budgets as needed. If our total budget for the year was determined to be $120,000, then we would create a monthly budget for $12,000 to track against. We know that if we exceed the spend in a month, then it could jeopardize our overall budget, and we would need to come in under budget for another month. Or if we were under budget for a month, we would have some flexibility on other months.

Once you define what your budget looks like, you can then create the alerting and action response for the budget. Figure 4-5 shows an example of a response.

Figure 4-5. *Example of creating a budget response*

Here, you will notice that you can create alert conditions when a certain percentage of your budget is met and select a corresponding action using built-in email alerting or action groups. You can also add multiple conditions with varying responses based on the severity of the threshold. For the lower thresholds, you might just send an email to inform people and then gradually increase the response each time it grows until it's adequately addressed. With the action groups, you have several different things you can do, and we will go through what each of those is in the next section.

Here is an example of a response on a monthly budget. Note that this is just an example and not necessarily something we are suggesting you set up. Ultimately, it is up to you and your organization to determine appropriate responses for certain thresholds.

- At 80% of the budget, send the team an email letting them know.

- At 90% of the budget, send an email to the team and include a level of management.

- At 100%, an ITSM ticket is created and investigation is required.

- At 110%, leadership is emailed to be made aware.

- At 120%, an automation runbook is executed to drop resources to lower tiers.

- At 130%, an automation runbook is executed that starts turning resources off.

Budgets are also editable and can be adjusted as things change. Suppose you take on a new large client midyear or saw an unexpected increase in volume and you are granted additional room in the budget to accommodate that, then obviously you would want to make changes to these. Budgets are designed to assist you so there is a degree of management to them to make sure they are doing that.

In addition to the budget alerts themselves, you can also change the language the email is sent in. If you are working with international teams within the same tenant, then this could be something that you take advantage of.

When you are looking at Azure Cost Management + Billing in the portal, you will also see a section called cost alerts in the cost management menu. Cost alerts are a list of responses to budgets that have occurred and whether or not they are still active. From this screen, you can see anything that has triggered and use the hyperlinks to navigate for more information about or to investigate further. If you click to create a new alert from this screen, it takes you to the screen to create a budget.

When looking at cost analysis earlier in the chapter, we mentioned being able to view budgets against your cost data in the chart views. When navigating through your data in cost analysis, just select the drop-down under budgets and select the name of the budget to apply it to your view. It will show based on your current data points how you are trending toward that budget. The great thing about this is the budget amount changes based on your level of granularity. For example, if you have a $30,000 monthly budget created and you change the granularity to daily, then the budgeted amount for the day would change to $1,000.

Action Groups

When creating budgets, you have the options to trigger an action group. These are notification preferences or actions that can be triggered when an alert condition is met. These will be covered again later in the monitoring section of the book, but we think it is relevant to take a quick look here. Let's walk through what each of the available options is and what you can do with each of them:

- **Email:** Send an email to an individual, distribution list, or group of users.

- **Email Resource Manager Role:** Subscription to the primary email address of any individual subscription owners. Active directory groups or service providers are excluded.

- **SMS:** Text messaging

- **Voice:** Phone call

- **Automation Runbooks:** Execute an automation runbook to perform an administrative action. This includes graphical, PowerShell, and Python runbooks. Any activity you can execute through those languages is supported. We use these often to scale, provision, deprovision, or suspend/resume resources.

- **Azure Function:** Run a function app. Azure functions are a serverless compute service that allows you to run code without the need to provision hardware. This supports a variety of coding languages including C#, F#, JavaScript, Python, and Java.

- **Webhook:** Deploy a webhook. This is an addressable HTTP/HTTPS endpoint used to communicate or post with other applications. These could be used to communicate with another application to trigger something or could be used for something as writing to a Slack channel.

- **Logic App:** Run a logic app. Logic apps are integration tools that help you schedule, automate, and orchestrate tasks across your organization. They can be used to perform a variety of actions.

- **ITSM:** Create a work item in your ticketing system. This supports connection to a variety of ITSM tools such as System Center Service Manager, ServiceNow, Provance, and Cherwell. If you are using a different tool, then check the official documentation to see if it's supported.

A lot of the actions can be based on preference and the skills associated with your team. With several of the services, you could actually create the same action, so it's a preference of what you prefer to use. But within these, the possibility of what you can do is limitless.

Note   Action groups provide a wide range of possibilities. If you can code it, you can do it.

Azure Advisor

The Azure Advisor is a tool that provides recommendations on how to best optimize your Azure deployments. Think of it as your personalized cloud consultant. Figure 4-6 shows an example of what those recommendations might look like.

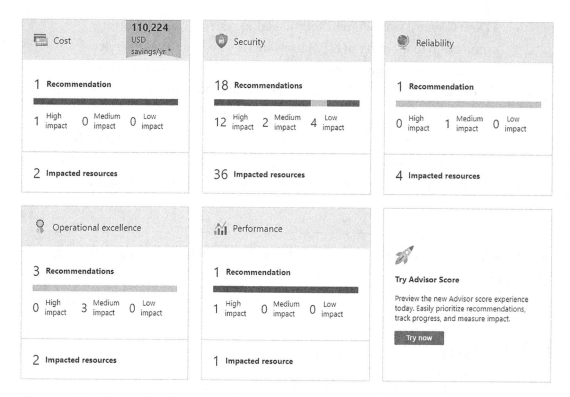

Figure 4-6. *Example of Azure Advisor recommendations*

While we are going to focus on cost optimization recommendations, the advisor actually covers a much wider range than just cost. It will analyze resource configuration and usage data to assist you in improving performance, reliability (formerly referred to as High Availability), and security. Ultimately, it provides you with the capability to be proactive with your environment and resolve issues before they spiral into bigger ones. We recommend checking this tool periodically for all areas to see if there are areas of improvement.

Here are some of the most common areas related to costs that the Azure Advisor may recommend, although we expect that this list will continue to grow:

- Resizing or shutting down underutilized virtual machines

- Right-sizing database server resources or virtual machines

- Reconfiguring or removing idle resources

- Recommendations for Azure Reservations

- Removing Azure Data Factory pipelines that are failing

- Storage type evaluation

To access your Advisor recommendations, you can access them directly from Azure Cost Management or by going directly to Azure Advisor itself. To access Azure Advisor, just navigate the Azure search bar and type "Advisor" and click to enter. If you access it directly from Azure Cost Management, you will only see the cost recommendations. However, if you access it from the Advisor screen, you will see all Advisor recommendations.

Just like other recommendation tools, you don't want to just blindly accept results. The tool cannot possibly account for every situation and variable, so the human element is really important here. Remember, they are recommendations and it's up to you to decide whether they are appropriate to implement. Take the time to make sure you understand what is being recommended and what the implications behind doing so would mean. For example, you wouldn't want to blindly accept an Azure Reservation since it represents a monetary commitment to Microsoft. Even though it is generally pretty good about making suggestions (especially with virtual machines), it does not mean it will always be. If you were to blindly accept the recommendation to create a reservation, it could mean significant financial penalties for your organization down the road if the assumption was incorrect.

Note   Don't just accept the recommendations. Take the time to make sure you understand what is being recommended and what the implications of doing so are.

Multicloud Tool

With the evolving cloud landscape, we are seeing more and more customers running in multiple cloud environments, so it's necessary that tools be able to support this. Azure Cost Management in the portal allows this by giving you the ability to monitor your Amazon Web Services (AWS) and eventually Google Cloud Provider (GCP) spend from within Azure. This provides you the ability to monitor all of your cloud spend collectively and from a single pane of glass. There is some setup involved in this process, and we recommend reading through the latest official Microsoft documentation on how to do so.

Azure Cost Management for Azure is free for any users under a supported agreement; however, for other cloud providers, you will be charged for using this. Currently, the cost is 1% of your total spend. So if you are using Azure Cost Management to monitor your AWS spend of $10,000 per month, then you would incur a $100 Azure charge each month.

Connecting External Sources

Microsoft does make the cost management data openly available if you want to access your cost management in other ways and build your own reporting. Most notably, if you are an existing Power BI customer, there is a connector that gives you access to your cost management data as long as you are under a Microsoft Customer Agreement or Enterprise Agreement. Previously, this was referred to as the Azure Consumption Insight connector and is now referred to as the Azure Cost Management connector.

To access the data is easy. Just open PowerBI desktop and go to the home ribbon. From there, select Get Data and then Azure Cost Management, as highlighted in Figure 4-7. You will need your Billing Profile ID or Enrollment number to be able to connect, as well as an administrator account or an account with permissions at the billing account or profile level.

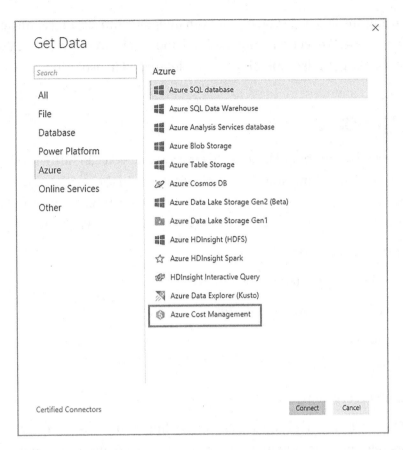

Figure 4-7. *Connecting PowerBI desktop to Azure Cost Management data*

For specific details on setting this up, you can refer to the official Microsoft documentation: https://docs.microsoft.com/en-us/power-bi/connect-data/ desktop-connect-azure-cost-management.

Once connected, you have access to pricing and billing data that you can then use to begin creating custom reports. Customization is a great benefit that a lot of people like to take advantage of, but utilizing external sources also allows you to create datasets that you can also combine with internal datasets. For instance, if you wanted to tie Azure billing data to another financial data source for more enhanced reporting, then you could.

For Azure Enterprise customers that want to build their own applications or are not Power BI customers, there are Cost Management APIs that provide the ability to access cost and billing data for analysis. This opens up the possibilities to be able to develop an application using any language you like or even integrate into existing applications you may already be using. We recommend reviewing the official Microsoft documentation on the details of this if this is something you are interested in exploring.

Lowering Costs

As you analyze costs, you should be looking for ways that you can further decrease your costs to make sure you are maximizing your investment. In the previous chapter, we looked at several programs and techniques to reduce costs. While we covered them at length there, they are also very relevant to keep in mind as you are managing your costs.

- Azure Hybrid Benefit

- Reevaluating architectural decisions

- Scheduled shutdown and startup

- Autoscaling

- Dev/Test pricing

- Azure Reservations

Some of these like Azure Reservations, autoscaling, and suspend/resume may become more relevant post migration since you might need more data points before making decisions on those.

Outside of those, a lot of cost optimization can come from analyzing your costs over time and tracking your spend to a budget. Maintaining a spend that comes under budget is the first component, and then the next component is looking at ways to further decrease that. Here are some additional ways you cut costs.

Look for underutilized resources that may be running with more resource power than required. This is not something that you are going to see from Azure Cost Management, but something that is going to require you to look at your performance monitoring tools or Azure performance metrics to evaluate. Remember that you are paying for what you consume so you want to utilize as much of the provisioned resources as possible.

Proactively spend time performance tuning so you get the most from your allocated resources. Spending the time tuning resources can have a significant impact on the underlying cost. Share with your teams the importance of doing this proactively and provide them with the time to be able to do so.

Look for waste and remove it. It is easy to spin something up in the cloud and then forget about it or deprovision a resource only partially. Azure Virtual Machines are an example of somewhere we see this often. Someone deprovisions the Virtual Machine but doesn't realize that they also need to delete the storage. Perform routine audits of your resources to make sure you need everything that you are paying for. You would be amazed at the amount of wasted dollars organizations spend on unused resources.

Look at access and determine if the users who have the ability to create and modify resources should have that access. The cloud provides us with a ton of flexibility to enable our teams to be able to do things quickly; however, it doesn't mean that it should become the wild west. You still need guardrails and need to make sure that those who have the power to create and modify resources understand the impacts of decisions that are made. Limiting access and holding people accountable is a way you can do this. Within most organizations, these should be driven by policies.

Engage with teams that are responsible for unknown charges. If teams are responsible for unknown charges, then open dialogue with them. It's possible that they are not even aware of the financial impacts of the decisions they are making. Creating accountability and having conversations can go a long way to mitigating overspend.

Create policies that enforce good governance on your organization. It's often not enough to create standards and documents and try to get people to adhere to them. While they always start out as a good idea, they always seem to steer off course at some point. Azure Policies can be used to enforce these standards to make sure your governance strategy is unfaltered.

Create good alerting that assists with being able to capture sudden changes in costs and unexpected spikes. You need to have an idea of what your expected cost is before you start deploying services. Take this information and create budget alerts at the beginning of your project so you can stay on top of things well before you get a surprising invoice. As you evolve throughout your cloud journey, you can get more specific with your budgets to be able to create granular alerting.

Utilize cost analysis often to evaluate and look for anomalies. Take advantage of this free and amazing tool to analyze your costs often and look for odd behavior. As authors, we cannot begin to speak about the number of times we have caught that has saved our organization significant dollars.

Work with Microsoft support to understand unknown charges. If there is ever a charge that you are unsure or doesn't make sense, then open a case with Microsoft to pose the question. Remember, there is a lot of complicated code frequently changing that runs behind the scenes to generate your bill and mistakes can happen. Earlier this year, we were using a service that resulted in a significant cost spike that didn't seem right, and we worked with Microsoft support to discover it was related to a bug. We ended up with a significant credit back to our account as a result. We think this is rare to occur, but it does occur. So if you ever see something that doesn't make sense, then it's worthwhile to go ahead and submit a ticket to determine why. Worst case, you get a better understanding of why something was billed a certain way, and it equips you better to make decisions on that resource going forward.

Summary

In this chapter, we continued the conversation on cloud cost management by looking at how to manage your costs in the cloud. Every good cost management strategy starts with planning, and this means we do our budgeting due diligence and have a solid plan for our Azure Governance framework. Microsoft provides us with an amazing toolset in Azure Cost Management + Billing in the portal to be able to manage our costs. This toolset helps create more visibility by providing access to our cost data using advanced analytic reporting in cost analysis. We are able to create more accountability in our organizations by understanding who is responsible for changes that impact cost. We also have budgets and alerting with automated responses that assists us during our busy days to stay ahead of anything abnormal. And lastly, we are able to better optimize our spending through Azure Advisor recommendations and several tips and techniques provided throughout this chapter and take advantage of the many cost-saving programs that Microsoft offers.

CHAPTER 5

Service and Systems Monitoring

Monitoring tools and processes have evolved in significant ways as the platforms they monitor have moved from physical, on-premises environments to fully cloud-native environments running Platform as a Service (PaaS) offerings and taking advantage of cloud features such as Autoscaling and Availability Services. Per-server monitoring has given way to monitoring systems and end-to-end Application Performance Management (APM) tools. Entire Site Reliability teams are dedicated to ensuring that data platforms and the business applications they support are performant and available for end users and customers around the world.

In this chapter, we will review best practices for monitoring Microsoft Data Platforms, including what metrics to collect from your on-premises platforms to help make your move to Microsoft Azure easier (and less expensive). We will discuss observability and how new technologies are changing the way data professionals detect and respond to issues and events within their environments. Additionally, we will review performance baselines, which we may sometimes refer to synonymously as benchmarks, and why they are important. Finally, we will demonstrate how to use monitoring data to reduce risk and costs for your Azure Data Platform migration.

Monitoring and Observability

A friend once told me that ice fishing is like a hunter looking up through a chimney waiting for a duck to fly over (he is Canadian, naturally). Unfortunately, many organizations still take this approach to monitoring their critical platforms. They use highly specialized tools, monitoring for extremely specific event occurrences, and then miss the larger picture of their organization's overall health and availability.

© Kevin Kline, Denis McDowell, Dustin Dorsey, Matt Gordon 2022
K. Kline et al., *Pro Database Migration to Azure*, https://doi.org/10.1007/978-1-4842-8230-4_5

Monitoring has historically meant collecting data and reporting on system and application availability and performance. This typically means a monitoring application or platform collects data from monitored targets through agents installed on the targets via software "polling" monitored targets. Other monitoring applications avoid the use of agents in favor of a "monitoring" server, which issues queries directly against the targets to collect various types of performance data using scripting languages like WMI or PowerShell or, for relational database monitoring targets, SQL queries against system tables.

The landscape of commercial monitoring tools and utilities is a competitive one, and there are a large number of choices for monitoring your data platform. Choices range from highly specialized, dedicated data platform monitoring tools like SolarWinds' SQL Sentry and Red Gate's SQL Monitor to broad APM products performing end-to-end monitoring, such as Dynatrace and Datadog. Additionally, the expanding world of open source and crowdsourced software offers a host of useful monitoring solutions, such as Nagios, and a large number of monitoring scripts publicly available on GitHub. In practice, organizations typically use a combination of specialized and generalized tools either by design or through acquisition.

The goal of a monitoring platform is to provide visibility into the health, performance, and availability of the monitored systems. This data is used in incident and event notifications and response, capacity trending and analysis, alerting, and ensuring that system performance meets or exceeds that which is required to support critical business applications and processes. Many organizational SLAs describe performance thresholds that system administrators and database administrators must measure against the output from their monitoring application. Without the performance information provided by a monitoring application, it's impossible for system administrators to know if their teams are meeting their contractual obligations.

Observability Platforms

Observability platforms moved to the forefront ahead of monitoring platforms after the release of Google's famous book, *Site Reliability Engineering,* available online at `https://sre.google/books`. Conceptually, observability is a superset of features found in monitoring. Its adoption is driven by organizations who need to broaden the capabilities of monitoring to represent a 360-degree view of systems and platforms in a rapidly changing cloud environment. In fact, job titles like Site Reliability Engineer and Database Reliability Engineer are now widespread. These roles are dedicated almost exclusively to ensuring sites and applications offer high performance and high availability, and are nearly ubiquitous within modern cloud-centric IT organizations.

Observability platforms solve the difficult choice technical leaders often face between using separate monitoring tools for individual components of their infrastructure stack and using a single-pane-of-glass dashboards by presenting business and technical metrics from multiple tools in a single, customizable, interface. They do this by ingesting data from various monitoring components via API, storing the data in a database, and then presenting the relevant performance data via a unified interface. The result is a single interface with customizable dashboards that are specialized for different personas within the IT enterprise.

Technical stakeholders are able to create dashboards with information related to performance and availability, while business stakeholders or technical leaders are able to track SLA compliance and customer metrics. Figure 5-1 shows how a collection of different tools ingest data from different sources to build the various dashboards the business needs. Many times, enterprises acquire a variety of monitoring tools in a piecemeal fashion as their various silos come to better understand their requirements and obtain budget dollars. However, the very fact that these monitoring tools do not integrate with each other means that the enterprise is unable to achieve a 360-degree view of their internal IT operations.

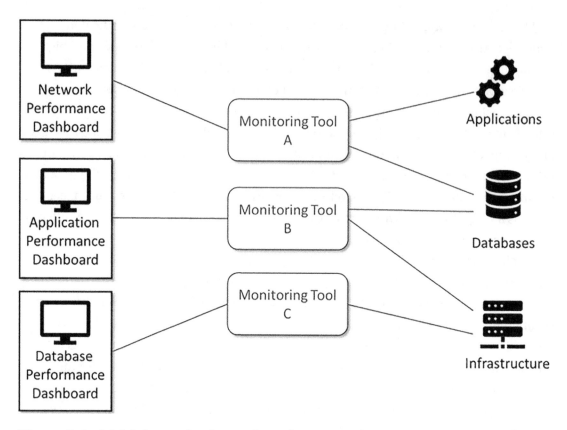

Figure 5-1. *Multiple monitoring tools and user interfaces without observability*

By comparison, Figure 5-2 shows the advantages that an integrated observability platform provides. The advantage – which you can see in the figure – is that all the results are combined into a single, unified dashboard that ensures that all executives are seeing and acting on the same information.

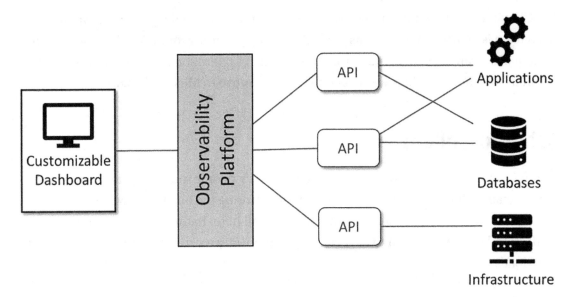

Figure 5-2. *Using API integrations and observability platform to create a single pane of glass*

We supported a large financial services organization who, over time, had grown largely by acquisition. As a result, they used multiple dedicated monitoring tools to monitor different parts of their environment. Networking, database, systems, hypervisor, and security all used different tools, providing reports to management which required multiple staff to spend hours every month to export historical data to CSV files and then process that data using Microsoft Excel.

Moreover, when an incident or event occurred, nobody had a single view of the impact or potential root cause, which in turn served to extend the duration of the incidents. By aggregating the data from multiple tools into a custom observability platform (and collapsing redundant tooling into fewer tools), users are now able to view metrics in customizable dashboards that are relevant and directly related to their roles with little manual effort. Moreover, the Site Reliability Team was able to identify correlations between seemingly disparate user processes and system metrics. A correlation between a specific workflow, an increase in the runtime of a scheduled stored procedure, and a spike in latency and errors in user connections at the front end was used to create automated remediation and notification before the issue affected users. The correlation was sent to the development team to optimize the code and resolve the issue in a permanent way.

Changing your organization's perspective from one of siloed monitoring, where each team is viewing their systems through chimneys waiting for incidents to occur, into system observability can be transformational to an IT operations organization and will result in substantial savings when it comes time to move to Microsoft Azure.

What to Collect

There are numerous blogs on the Internet, written by people much smarter than yours truly, detailing the metrics that are critical to monitoring the health of a Microsoft SQL Server data platform, so we will not list them all here. Rather, it is important to understand the *types* of data to collect and the importance of understanding end-to-end workloads.

Metrics, Logs, and Traces

Broadly, information collected by monitoring and observability platforms can be broken down to *metrics, logs, and traces. Metrics* are used to identify trends and send alerts. One example of metrics is the logical and physical disk telemetry produced by Windows Performance Monitor. *Logs* are event-based and verbose chronological records that provide detailed information about what is happening within a system at a specific point in time. An example of a log is the Windows System Events log. *Traces* are request based and track a specific activity or process, for example, the default system health trace available by default in Microsoft SQL Server. In combination, all of these contribute to a 360-degree view of platform health and availability.

The type of data required to effectively catch and diagnose issues depends on where in the application stack you are looking. For instance, front-end application server health monitoring relies almost entirely on metrics generated by real-time user monitoring (RUM) or synthetic transactions, while back-end database systems are monitored using metrics and log data.

Metrics measure how a system is functioning in a very specific way at a specific point in time, such as CPU Utilization/Sec or Database Transactions/Sec. Metrics can be broken down into two categories – Work Metrics and Resource Metrics. Work Metrics are indicators of overall health of a system as indicated through their output. They measure such things as throughput, success, errors, and performance. Specific thresholds may be defined, which, when exceeded, generate an alert or trigger a remediation action.

Resource Metrics measure a system's state via utilization and saturation. For instance, disk saturation expressed through queuing or waiting database transactions are examples of resource metrics. Table 5-1 details common Work and Resource types and subtypes and common workloads captured for each.

Table 5-1. *Work and resource metric examples*

Type:Subtype	Name
Work:Database Throughput	Batch Req./sec
Work:Error	% requests > SLA response in seconds
Work:Performance	90th pct querytime in sec
Resource:Disk IO	Disk queue length
Resource:Memory	% memory in use
Resource:Application	Queued session requests

Work and Resource Metrics, when viewed as a whole, provide a holistic view of your systems' workloads. A comprehensive understanding of a system's workload is key to using observability metrics to inform your move to Azure.

All too often, administrators and architects focus on specific metrics related to health and availability without understanding the overall workload. Poor decisions often follow. For example, they often adopt the "lift-and-shift" methodology of building their Azure landing zone with the same compute and storage resources as the on-premises systems they are meant to replace. This results in excessive costs by paying for unused or unneeded capacity (refer to Chapter 3 for more details on this issue). Consider these questions:

- Is your systems' resource utilization consistent, or are there periodic usage peaks or troughs (e.g., for reporting periods)?

- What percentage of resources is currently being utilized on-premises?

- How much headroom has been purchased for growth?

- Transactional or analytic workload?

- Is the data workload primarily compute or storage intensive?

Administrators must be able to qualitatively answer all of these questions (and more) to ensure that they are neither spending too much by overprovisioning resources in Azure nor short-changing their users on performance by underprovisioning resources in Azure.

From Servers to Services

It is also helpful to switch your perspective from *server* to *service* availability. All of the individual servers and virtual machines we manage are part of larger applications, stacks, and systems. The resources comprising these systems are not of equal importance to the enterprise and need not be treated as such. Individual resources can be assigned weighted scores according to their relative impact to the application system or service, with each contributing to the overall health score of the service. Your aggregate health score is then calculated based upon the health scores of each component. In doing so, the appropriate thresholds and incident response workflows may be configured for each. Monitoring collections of resources as services provides a more meaningful look at the health and availability of that system and its components holistically.

For instance, consider a three-tier web application such as the one shown in Figure 5-3 with five load-balanced web front-end servers, two middle-tier servers performing business functions, and a three-node SQL Server AlwaysOn Availability Group. Using a weighting of 1 (low) to 5 (high), each web server is assigned a weight of 2; the middle-tier servers have scores of 3 and 4. The primary SQL Server has a weight of 5, while secondary nodes are assigned weights of 2.

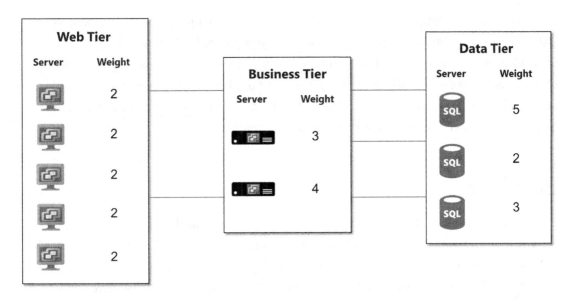

Figure 5-3. *Three-tier application with assigned weights*

In the scenario shown in Figure 5-3, a single web front end experiencing issues is not as impactful to overall system performance as a disk issue on the primary SQL Server node. The overall score of the application service is an aggregate of the scores of each component. By viewing the resources as part of an application system and assigning weights to each, the appropriate action may be assigned to a triggering event based upon overall impact to the application's availability.

Another important advantage of monitoring services is the ability to design and implement predictive monitoring. Many organizations we've worked with are entirely reactive in their incident response. For them, the tyranny of the urgent, fire fighting, prevents them from doing what's important, such as adding value to enterprise IT processes and workflows. In situations like this, we often ask them if they are paying their customers to be their application monitoring service. After all, an IT organization has failed at a core component of their mission if they are being informed of an issue by a customer or end user. It happens, we know – but it should only happen once.

Note   If you are in the habit of relying on calls from users to determine whether your systems are performing well, you're doing it all wrong.

Service monitoring is as much a mindset as it is a set of processes and workflows. Using service monitoring and observability, IT teams are able to identify precursors to service-impacting events. These "canary in a coal mine" conditions permit teams to proactively take action when an issue is imminent, or to inform their customers of an issue before it becomes critical. This instills far greater customer confidence in the organization and helps build your credibility as a team.

Baselines

Establishing and using baseline values for key performance and availability metrics are key to understanding your production workloads. Additionally, effective baselines help to properly size your Azure environment using real data from your production systems. Rather than simply building in Azure what you have on-premises, we can use our baseline metrics to determine what capacity is truly needed and whether we should leverage advanced cloud features like autoscaling and elastic pools rather than paying for unnecessary and unutilized capacity. Moreover, baselines can play an important role in reducing noise (e.g., non-actionable events and alerts) by creating smart alert thresholds rather than using fixed values.

Baselines vs. Thresholds

The terms "Baselines" and "Thresholds" are often used interchangeably, but there are significant differences in their meanings. Thresholds describe a specific level of resource consumption or workload behavior specific to a known timeframe. It is a level at which something can be measured or judged, and actions taken in response. Baselines, on the other hand, are used to define the typical state of a system.

Baselines are averaged over a period of time and identify usage patterns. In the presentation Baselining and Alerting for Microsoft SQL Server (`www.sentryone.com/white-papers/performance-baselines-and-benchmarks-for-microsoft-sql-server`), baselines are described as "lines in the sand." Establishing and using baselines in routine service reviews, for capacity planning, and for smart alerting are signs of a mature IT organization. Another important term, benchmarks, describes the collection of baseline values under a predetermined level of stress and workload, for example, the industry standard TPC-C and TPC-H database performance benchmarks.

Note   A **baseline** defines what is normal for a given system and can be used as a measurement against which future performance is compared. A **threshold** is the value at which an action is triggered. A benchmark is a baseline used to collect performance metrics under a predetermined workload.

IT professionals create mini-baselines every time they troubleshoot a performance issue. In fact, effective troubleshooting is virtually impossible without them. One of the first steps in diagnosing a performance issue is to establish what "normal" is – what was happening on the system during normal and unremarkable periods of operation when the systems are operating properly and users are happy with system performance? During troubleshooting, we must collect performance metrics immediately before and during the time period the issue occurred and determine the scope of the issue. What metrics spiked? What changed? These are questions directly related to performance baselines.

Note   If you have not spent time creating a baseline to determine the value of "normal" performance metrics, how can you tell when those metrics are abnormal?

Figure 5-4 shows the use of baselines in the process of troubleshooting performance problems. You can see the sequence that begins with looking at a baseline and then at what metrics changed or became abnormal, and ultimately that leads to a solution that is tested and implemented to ensure the abnormal situation or incident is avoided in the future. If the changes are successful, then the IT team makes the changes permanent and possibly creates or updates the existing alerts to ensure that similar incidents in the future are predicted even earlier in the service monitoring process.

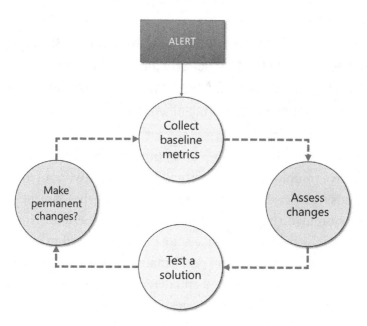

Figure 5-4.   Using baselines in performance troubleshooting

It is important to create baselines on the correct metrics. While there are certain standard metrics that are almost always included in baselines, such as the ubiquitous metrics CPU, Disk IO, and memory utilization we find on every Windows laptop and desktop. However, there are probably many other business-specific metrics that must be included to ensure effective service monitoring and provisioning of Microsoft Azure resources.

For instance, a company we worked with had daily scheduled operations during which text files were read and loaded into a Microsoft SQL Server database. Once the data loading operation was complete, a series of SQL Agent jobs were executed on a schedule to create payments for customers. The number of payments were normally numbered in the tens of thousands per day, and the financial penalties for a late payment were substantial. Moving the platform to Microsoft Azure included not only the SQL Server instances but also enterprise job schedulers, scheduled tasks, and complex SSIS operations.

To ensure the solution would be able to support the business requirements, we determined what metrics best measured the throughput required to process the payments daily. Additionally, we identified the conditions that created bottlenecks causing missed payments. Where appropriate, we optimized SQL and SSIS code to run more efficiently. We also established meaningful custom baselines for these metrics and used those baselines to inform the sizing for the Azure environment. The move to Azure was a success, and missed payments later dropped as a percent of total payments.

The service Azure Application Insights also has the ability to build smart baselines using historical performance data. The baselines produced by Application Insights may then be used to build dynamic thresholds and smart alerts, which provide more meaningful and actionable alerts. Since workloads change over time, the ability to adjust monitoring baselines and thresholds dynamically ensures that critical conditions are not missed while at the same time reducing alerting "noise" generated by transitive situations or alerts for which no action is required.

Using Baselines to Build Smart Thresholds

A key objective of any operations organization is to reduce noise caused by non-actionable alerts. The principle in play here is that if an alert is raised for a particular threshold, it must be actionable. If it is not actionable, it can be logged or sent to an email inbox to be used for informational, trending, or forensic purposes. (We've never actually GOTTEN to 100% actionable alerts, but we've worked with many teams over the years to dramatically reduce the number of tickets created for minor conditions.)

Traditional monitoring notifications are triggered based upon fixed values for a given threshold. For instance, an alert is fired sending administrators an email notification when CPU utilization reaches 85%. Another example for SQL Server might be when Log Disk capacity drops below 15%. In the cloud, however, you pay for what you provision – not what you use – and are able to size resources up and down dynamically. Where our on-premises systems averaged 40% utilization, our Azure resources may average 70% utilization to avoid overspending. We may also use thresholds to trigger autoscale events to add capacity when traffic spikes or an application server goes down. This requires giving additional consideration to alerting and action thresholds and to the use of performance baselines to detect anomalies.

Note   Dynamic alerting thresholds are an important way to reduce non-actionable alerts ("noise") in cloud environments, where resources tend to run at higher utilization levels than they do in on-premises environments.

Anomaly-based alerting using established baselines eliminates the noise created by fixed thresholds by creating alerts based upon deviation from baseline. For instance, an enterprise may have a production threshold of 95% for CPU utilization at which an alert is sent to the IT team. It is not uncommon for this alert to wake the on-call person in the middle of the night during server maintenance or other normal operations. They could then put a maintenance window in place during which notifications are silenced, but they then run the risk of missing a real emergency. Additionally, they've migrated the systems to Microsoft Azure, and because they are a well-run organization who read this book, they've right-sized their Azure resources they run at sustained CPU utilization values of 80% or more.

Rather than using the fixed 95% threshold, what if they were to create an alert that fired when CPU utilization exceeded the baseline by 20% and stayed at or above that level for 30 minutes? This would allow for temporary spikes due to maintenance and would ensure that transitory conditions did not (blessedly) result in alerts to IT personnel. Because they are a well-run organization, they set the 30-minute duration to be below the service-level agreement (SLA) with their customers. The alert threshold would also move with the baseline so that if a code update results in lower consumption, the threshold would lower with the baseline. Finally, using the correct timeframe for thresholds could accommodate those pesky month-end reporting spikes that may exceed the 30-minute value.

Not all monitoring and observability tools support the establishment of baselines, but it is becoming a standard feature found in most tools. Creating baselines and thresholds may require that custom or nonstandard metrics are defined. Many tools are quite sophisticated in this regard and support custom baselines that permit organizations to define workload-specific baselines. These baselines are a powerful tool in that they reduce alerting "noise" resulting from non-actionable notifications and enable those teams to assume a proactive stance on system performance and availability.

Some monitoring and observability tools enable the creation of thresholds that are a calculation, usually a baseline metric calculated against a statistical function, which enables even more sophisticated smart alerting. Let's take our earlier example of 80% CPU utilization as a key threshold. In this example, any time our system sustains a load of 80% CPU busy for more than 30 minutes or more, the alert will fire. Clearly in that case, the system is experiencing abnormally busy CPU issues, possibly resulting in a slow user experience. However, what about situations where the system suddenly and unexpectedly drops to only 10% CPU utilization – on a system that's never seen less than 30% utilization? Perhaps a major segment of our network is offline, meaning that a large number of users simply cannot connect to the system. That's clearly an abnormal performance metric, but our threshold is only set to alert us when CPU consumption goes too high. In a situation like this, you might wish to create a threshold based on the mean value CPU utilization times 3 STDEV on a normal distribution, meaning that you would receive an alarm whenever CPU rises 15.8% above or 15.8% below the mean. So with one simple calculated threshold, you can even further refine your baseline threshold values.

Using Baselines to Right-Size Your Cloud Platform

In the cloud, you pay for what you provision, not what you use. We have counseled clients in some cases that it is better for an organization to wait until they are ready to migrate *into* their cloud-first future state, which we've referred to as a *landing zone* throughout this book, with containers and microservices than it is to try to lift and shift into an IaaS solution first. However, the "lift-and-shift" methodology is often the best first step for an organization, enabling some though not all of the cost savings and efficiencies of the cloud by using the Infrastructure as a Service (IaaS) architecture. The ability to right-size workloads on VMs in the cloud, instead of buying and overprovisioning on-premises servers for future growth, may justify the move to virtual machines in Azure.

Baselines are an invaluable tool when sizing your Microsoft Azure Data Platform. Whether migrating VMs to an IaaS solution or migrating to Managed Instance, Azure SQL Database, or another PaaS offering, you will be prompted to select sizing for the target resources. Oftentimes, organizations overprovision resources to be "on the safe side." This is often due to organizations not having a detailed understanding of the workloads running on their infrastructure. In many cases, the intent is to temporarily overprovision and at some point downsize as necessary after migrating. Our experience is that the overprovisioning becomes permanent and the company runs through their cloud budget more quickly than expected because they are paying for unused capacity.

This is usually where a team is created to identify cost-savings opportunities, although the team is sometimes composed of people who have a vested interest in NOT downsizing the resources they manage. This can lead to acrimony between teams and otherwise cause intrapersonal friction that is unhealthy for the IT operations team.

Baselines are your bellwether. When determining how to size your Azure Data Platform resources, where companies normally used anywhere from 40% to 60% sustained utilization as the threshold at which they increased capacity (e.g., bought hardware) on-premises, we are now able to dynamically scale up (or down) elastically. So running at 70% or higher sustained utilization is not uncommon. You can also enable autoscale events to trigger scaling out resources when utilization reaches a certain threshold, or scheduled to accommodate periodic workload requirements.

We have worked with many organizations on reducing their cloud spend. There are a number of important things enterprises can do to save money. Automatically shutting down nonproduction resources at night and on weekends, tiering storage for hot/cold/archive, and autoscaling are all effective ways to reduce cloud costs. However, none of those methods come anywhere close, in pure dollar savings, than an understanding of production workloads and creating effective baselines to right-size their cloud resources. When used in conjunction with Reservations and Reserved Instances as discussed in previous chapters, significant savings may be realized through benchmarking or your data workloads.

Data Platform Monitoring in Microsoft Azure

Of course, the work does not stop once you've moved to Azure. In fact, tighter SLAs and less performance headroom coupled with new technologies means data professionals must be even more in tune with their workloads. Additionally, many of the parameters used to tune performance on-premises are abstracted in Azure PaaS offerings. Sure, maybe it was a pain to manage Operating System (OS) settings in your virtual environment, but at least you could. Admins now have fewer sliders with which to tune performance.

Cloud-native monitoring is an area of tremendous innovation. In fact, if asked where we think there is likely to be major disruption within the next three years, we would say, without hesitation, monitoring and observability in the cloud. Streaming datasets, Machine Learning (ML) and Artificial Intelligence (AI), and OpenObservability are harbingers of what is to come. Virtually all commercial data platform monitoring tools now support monitoring Microsoft Azure resources in some fashion.

Azure Monitor

Initially, the monitoring tools provided through Microsoft Azure were limited and largely used to supplement other tools. That is changing as new features are added to Microsoft Azure Monitor at a rapid clip. Azure Monitor provides the ability to gain a unified view of your on-premises and Azure resources with cloud-aware monitoring to gain deep insights into dependencies, application performance and availability, and key metrics.

Azure Monitor provides a dashboard view of your enterprise platforms and relevant information in customizable format. Azure Log Analytics gives admins the ability to create enormously detailed insights by querying logs for all resources within an environment. Additional tools are available to provide guidance on cost optimization and to assess risk in data and other platforms through best-practice analysis and recommendations. Figure 5-5 is an example of an Azure Monitor dashboard displaying critical information about the health of an environment.

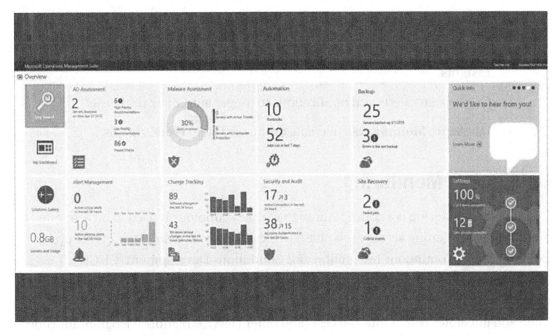

Figure 5-5. *Azure Monitor dashboard*

Azure monitor collects and stores metric and log data for monitored resources. As discussed earlier, metrics are numeric and measure one or more aspects of a system's performance at a particular point in time. Logs contain detailed records of different types of events and can be chock-full of verbal information. Using metrics and logs, Azure Monitor provides full-stack monitoring observability across applications and infrastructure for both cloud-native and on-premises systems. The Azure Monitor ecosystem includes numerous partner integrations with widely used Application Platform Monitoring (APM) tools and other ITSM platforms to provide a unified observability platform for your organization.

Note   Azure Monitor collects and stores metrics and logs. These records may then be used together for analysis of incidents and events.

Azure Monitor uses metrics and logs to perform a variety of functions:

- Use **VM Insights** and **Container Insights** to correlate infrastructure issues with application incidents and events.

- Use **Log Analytics** to perform deep analytics of your monitoring data.

- Identify issues across application dependencies with **Application Insights**.

- Create smart alerts and notifications to trigger automatic responses.

- Use **Azure Monitor** data to create dashboards and visualizations.

Continuous Monitoring

Continuous monitoring is a concept through which technology is used to incorporate monitoring and alerting across each phase of the IT operations and DevOps life cycle. Combined with Continuous Integration and Continuous Development (CI/CD), continuous monitoring helps organizations deliver solutions more quickly and reliably while ensuring optimal performance. Azure Monitor works using Visual Studio or VS Code and integrates with Azure DevOps (and other DevOps platforms) to monitor code deployment and operations.

For the data professional, continuous monitoring provides a means to detect issues with application or database changes, to trigger notifications, or to enable automated rollbacks. For a long time, database code was handled separately from application code in CI/CD release pipelines. It was (and, in many cases, still is) not uncommon for application code to be rolled out via DevOps deployment pipelines, while DBAs run T-SQL scripts for corresponding database updates without CI/CD. This required manual execution and the inclusion of rollback scripts in the event of an issue with the update. As organizational processes mature and database updates are included in automated deployment pipelines, continuous monitoring has put data platforms on the same footing as application and infrastructure code.

Data Platform Monitoring Tools

Microsoft Azure contains several features specific to monitoring data platform components. Using the data collected by Azure Monitor as well as some specific features for Microsoft SQL Server or Azure SQL Database, database administrators and other data professionals are able to gain valuable insight into the overall health of the resources they manage.

SQL Server Health Check

Azure SQL Health Check is an agent-based solution used to assess the overall health of your SQL Server environment to provide scores and recommendations across five focus areas:

- Availability and Continuity
- Performance and Scalability
- Upgrade, Migration, Deployment
- Operations and Monitoring
- Change and Configuration Management

The agent collects data using Windows Management Interface (WMI) metrics, Windows Registry information, performance counters, and SQL Server Dynamic Management Views (DMVs) and then forwards it to Log Analytics every seven days. Upon completion of the assessment, detailed health scores and specific recommendations are provided via the SQL Health Check Dashboard as detailed in Figures 5-6 and 5-7.

Note   SQL Health Check uses WMI, Windows Registry, performance counters, and SQL Server DMVs and requires the agent to run under an elevated Run As account with the appropriate permissions.

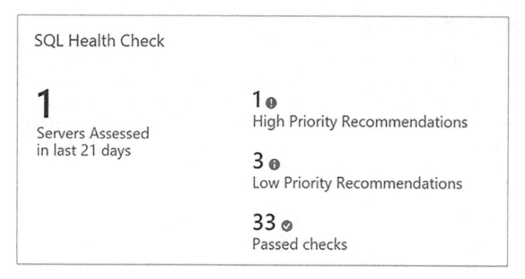

Figure 5-6. *SQL Health Check summary*

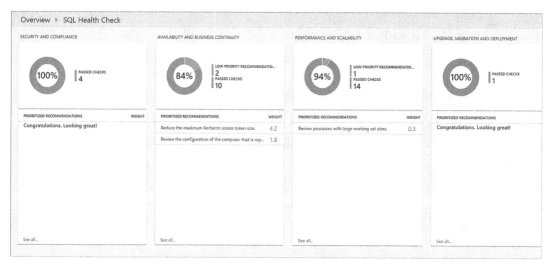

Figure 5-7. *SQL Health Check scores and recommendations*

The suggestions provided by the SQL Health Check are weighted using a relative importance for that recommendation. Weights are aggregated using three key criteria:

- The likelihood that an issue will cause a problem

- The anticipated impact to the environment if a problem occurs

- The level of effort required to implement the recommendation

It is not necessary to shoot for a 100% score for every environment or server as each environment is different and there may be very good reasons NOT to implement a recommendation. Additionally, recommendations should be evaluated in a test environment where appropriate. I'm sure many of you have encountered environments where a junior DBA applied all recommendations from the Database Tuning Advisor and spent a good deal of time evaluating indexes with the "_dta_" prefix. While the recommendations provided by the SQL Health Check tool come from a vast repository of knowledge and product understanding, it is important to evaluate how they will affect individual workloads and environments. However, good DBAs and others are already doing this analysis, in many cases using their own DMV scripts and manual assessment of historical performance data. SQL Health Check gives those same professionals an effective and time-saving tool in their platform optimization toolkit.

Azure SQL Analytics

In preview at the time of this writing, Azure SQL Analytics is a cloud-based monitoring and analytics tool for monitoring performance across all of your Azure SQL databases in a single dashboard. The tool collects metrics and allows administrators to create custom monitoring rules and alerts. Paired with Azure Monitor, data professionals are able to gain detailed insights into each layer of their application stack and quickly make correlations and identify issues when they occur. In order to use Azure SQL Analytics, data is stored in Log Analytics to take advantage of the dashboard and predictive analytics. There are costs associated with ingestion and storage into Log Analytics, but the results can be incredibly useful for DBAs and Admins and may allow you to consolidate some of your tools to offset the expenses.

Azure SQL Analytics is a relatively lightweight solution and does not support monitoring of on-premises SQL Server instances. Instead, it uses Diagnostic Settings to send Azure metric and log data to Log Analytics. It supports SQL Elastic Pools, Azure SQL Managed Instances, and Azure SQL Databases. Azure SQL Analytics provides detailed information on critical SQL Server performance metrics detailed in Table 5-2.

Table 5-2. *Critical SQL Server performance metrics*

Option	Description	Notes
Resource by Type	Counts all resources monitored	
Insights	Intelligent performance insights	
Errors	Hierarchical view of errors	
Timeouts	SQL Server timeouts	Not supported on Managed Instance
Blocking	Hierarchical view of SQL Blocks	Not supported on Managed Instance
SQL Waits	SQL Wait Statistics	Not supported on Managed Instance
Query Duration	Query execution statistics including CPU, Data IO, and Log IO	
Query Waits	Query waits by category	

Once configured, Azure SQL Analytics provides a dashboard view with drill-through charts to review performance details collected by the tool. Figure 5-8 provides an example of such a view.

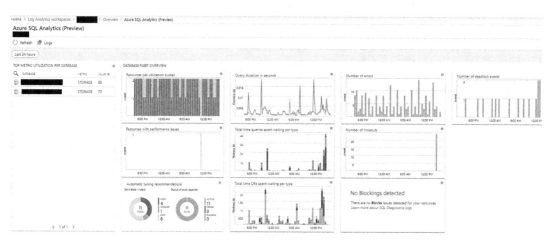

Figure 5-8. *Azure SQL Analytics dashboard view (Azure SQL DB)*

Summary

We once consulted with an organization who, like many others, struggled with containing costs. This was, in part, because of directives that drove bad behavior ("We are going to be x% in Azure by x date!") and a lack of understanding of the performance requirements and the difference between on-premises and cloud provisioning and sizing. Any suggestion of waiting until we were ready to move in a cost-effective manner was cast as that person being "anticloud." They hit their targets by building new resources in Azure without moving any production systems. They also spent their first year's Azure budget in seven months. People were fired because of it.

Monitoring and observability are most often discussed in terms of IT Service Management (ITSM) and Site Reliability and may not be something many Microsoft Data Professionals spend much time thinking about outside of query tuning and incident response. In reality, it is one of the most important aspects of the modern DBA's job due to the implications for planning and executing a strategy for moving to Microsoft Azure. The cloud is changing how data professionals work.

They are no longer able to simply focus on the ducks flying over their chimney – they have to have at least a basic understanding of business requirements, budgeting, site reliability, and other aspects they used to ignore. The one thing we emphasize more than almost any other is "KNOW YOUR WORKLOAD!" Monitoring and observability are key components before, during, and after a migration to Microsoft Azure.

CHAPTER 6

Migrating Data and Code

When you are migrating a database to the cloud, you will want to give careful consideration to the code and data you are migrating. As you are already aware, we pay for everything we use within the cloud, so migrating our technical debt or unneeded components is going to cost, sometimes quite a lot. In fact, it is not uncommon at all for the use of legacy data types, outdated T-SQL, or things like the use of cross-database queries to be the determining factor between migrating to a VM in the cloud (IaaS) or a PaaS offering like Azure SQL or Amazon RDS. That can have a large impact on the ultimate cost of the migration. Last but not least, we also need to be sure that the code we are moving is supported on the service we are migrating to and does not have compatibility issues or is needlessly wasteful with billable resources.

Technical debt is the concept of taking the quick and easy approach now over the right way, which might be more difficult or take longer time. It's often well-known to those involved that they are taking a shortcut, but it is justified by the fact that we will come back later and do it the right way when we have more time. However, the timing for that does not always seem to present itself, and that item just sits at the bottom of the backlog or disappears completely. After all, you have a working solution, and there are other things not working at all or needing to be done that you are being asked to focus on. And the more you kick the can down the road, the easier it gets to continue postponing code remediation and the less optimal your environment becomes.

We will all experience being part of or creating technical debt in our careers, and it is just part of the cycle; however, some of these things can be detrimental to our cloud migration. Items such as running less optimal code, disregard for performance tuning, continuing to run deprecated processes, and maintaining old data that is no longer needed are things that can negatively impact your migration. These things result in more resources needed, which ultimately leads to higher costs. In an on-premises environment, we know these are things we should be looking at often, but because the

© Kevin Kline, Denis McDowell, Dustin Dorsey, Matt Gordon 2022
K. Kline et al., *Pro Database Migration to Azure*, https://doi.org/10.1007/978-1-4842-8230-4_6

return on investment (ROI) is greater in other areas, they sit at the bottom of the backlog. After all, we know that we can cover these issues up with overprovisioned hardware that we have already purchased, so there is no immediate out-of-pocket cost.

Sometimes though your technical debt may not be the issue, and you may just be splitting or removing functionality as part of the migration. For example, you have an application database that you are only migrating part of the application for or you are using the migration as an opportunity to remove features or functionality. Since you are already down in the weeds and you want to migrate as optimal a database as possible, then this can be a good time to look at that. Everything may not be the result of technical debt and may be related to decisions you are making. Whatever the case may be, you will want to do your best to optimize your code and data footprint to make sure you get the best price on your services.

Unless you are running cloud virtual machines, then your source and target are not going to be exactly the same. For example, if you are moving a database running on an on-premises SQL instance to Azure SQL database, then you are going to find differences such as compatibility conflicts or unsupported features. Sometimes, these are non-impactful, but other times they can be enough of any issue to halt the migration. You will not want to leave these to chance whether they will cause you a problem or not and will be something that you will want to check.

In this chapter, we are going to dive into these items a lot deeper and walk through the process and tools available to make sure you are only migrating the things you need to. We will start by focusing on just moving meaningful data and then follow that up with moving meaningful code.

Migrate Meaningful Data

If there is any axiom that holds true in the cloud, it's "the less you do/store, the less it will cost." While that may seem obvious, it can be an easy thing to forget as you're planning all the shiny new data platform things you get to do in the cloud. If you require less compute and storage to execute your workload, you'll pay less. While there are a lot of ways to do that (and later in this chapter, we'll walk through some tips and tools particular to refining code), the simplest way to cut down on what you're migrating to the cloud and running in the cloud is to reduce the amount of data you are moving.

> **Note**   If you require less compute and storage to execute your workload, you'll pay less. Move only what you need.

Do you know what data you can safely delete? In some cases, this requires a thorough review by business analysts, data professionals, and everybody in between. In other cases, it is as simple as asking your lone DBA what can safely be deleted from the one giant database that they support. Most organizations fall somewhere within that continuum, though.

Archive the Unneeded

What guidance can we give the people that will be doing this review? Primarily, keep it simple. You do not need to archive or delete every single piece of extraneous data before a cloud migration can begin. Are there tables with "archive", "log", "audit", "old", "tmp", "temp", or similar in the name? Begin with those tables and review their data to see if they can be archived or removed completely. In my experience, many tables with terms like that in their name can be partially archived at worst, or completely removed at best. It is important to note here, however, that you may be migrating third-party apps where this is difficult or impossible. Make your best effort here, but do not damage functionality along the road to migration.

For example, if we execute the following query against the example **WideWorldImporters** database, we can see that there are 17 tables with a suffix of "archive":

```
SELECT *
FROM sys.all_objects
WHERE type = 'U'
        AND name LIKE '%archive%'
        - modify and rerun for other archival-style table names
ORDER BY name;
```

Using the preceding code, you can quickly identify tables that are already indicated as old or archival data. To avoid deleting data that might be important only once or twice per year, we recommend creating a cold storage schema with a name like "TBD" or

"ARCHIVE" and then moving any such tables to that schema using the statement ALTER SCHEMA … TRANSFER. You can safely drop those tables after testing shows that they are no longer needed.

Along similar lines, do you have large tables with data that has never been archived or examined for archival? Initiate a review of those to see if data before a certain timestamp, ID number, etc., can be removed entirely or placed in an archival table where something like compression or columnstore indexes could be used to reduce the space needed to store that data while preserving the transactional table's structure and indexes. You may want to consult with the business users of this very large table to see if you must retain data from its entire life span. If, for example, your very large table has 20 years of records stored within, but the business users only need or use the last X years of data, then you could potentially save quite a lot of storage expenses in the cloud.

A word of caution is in order here, however. Many industries are governed by regulatory requirements about the length of time where data needs to be accessible to an auditor or someone similar. There are often regulations surrounding the length of time in which the regulated data must be available for an audit, although cold storage in the cloud or related cold tier technologies may be the good fit for every bit of data you can archive. A quick example is the American law known as Sarbanes-Oxley (SOX), which requires financial data from publicly traded firms to be accessible for audits for seven years after the data was instantiated. We will review some scenarios along those lines in the following section.

As a final bit of summary, if you have an on-premises database that has been running for several years, there is a high likelihood you have data you can archive, remove, or delete, and that will save you money as your migration progresses. Migrations require a lot of planning and effort, so take advantage of easy wins like that!

Understand Business Requirements

Many organizations make a personnel mistake when putting together a project team for a data migration. Migration teams are often overloaded with technical personnel or, worse yet, composed entirely of technical personnel. As we've been discussing in this chapter, business users and analysts have an important role to play in analyzing exactly what data you will be migrating to the cloud and what form that data may take somewhere along the way.

If your shop is small enough where the single DBA or the small DBA team wears both technical and business hats, then this analysis process can be completed quickly. The vast majority of shops, however, operate at a scale far beyond where one or two people could confidently answer those questions. Since data estates of that size make up a very high percentage of the total, it is important to take a few moments here to focus on the purely business user-driven portion of most migration projects.

If you have already completed the initial analysis described earlier, your project has reached the point where the business users and analysts need to be interviewed to see if they can help you reduce the volume of data headed to the cloud. There are two important categories to consider here: Who should we speak to and what types of questions should we be asking those people?

Who should we speak to? An important part of laying the groundwork for a successful migration is identifying the business users, analysts, and owners for every application whose data you will be migrating. This should be tracked in your project management tool of choice or, failing that, something simple like an Excel spreadsheet.

If it is possible to get the application's "business team" together, that is the preferred format for the meeting because it loops everybody into the process early on and generally having the team together on a call helps smoke out any questions or thoughts they have about the supporting data for the application. It can also help bring to light any quirks related to the application's data and overall environment. Those may be significant to discuss within the migration team as the quirks are often technical in nature.

Note   An important part of laying the groundwork for a successful migration is identifying the business users, analysts, and owners for every application whose data you will be migrating.

This is not always a simple process, but our general experience has been that regular meetings with technical leadership, data teams, and help desk personnel can help you put together a fairly comprehensive list of people who own or care about these applications. Having these discussions early on can also help ensure other people are looped into the process who may have significant roles in operating and supporting the application but who the technical personnel are unaware of.

Once these preliminary discussions are out of the way, you can dig deeper into the data itself. While you may get some eye rolls when you announce that you will be working with this team to perform that type of analysis, the best motivating sales pitch to stay involved in this is typically some variation of "if we do this correctly and quickly, we'll save the company money and look good." This may not always work, but it's a good starting point. You're all working toward reducing costs overall, and that is often incentivized in the form of awards, bonuses, and the like. If you can keep this team's eye on the possible prizes, this helps a great deal!

Once the team has bought into their part of the process, it is time to dig into data and ask the right types of questions. A great place to begin that part of the discussion is asking if any feature has recently (or not so recently) been turned off or deprecated. If it has, what tables, stored procedures, etc., supported that feature? Can those be removed? If the answer is "they can't be removed but their performance doesn't matter any longer," it is important to note that because it expands your options for storing/accessing that data (and likely decreases the amount of money you will spend to do that). Cheaper, slower storage for that type of data will have a positive material impact on the project and your budget as well.

Obviously, if the feature has been removed and the data is no longer needed for any reason whatsoever, your options are quite simple. All you need to do is figure out the most efficient way to delete it, ensure you have one last archive of it, and remove it. Bear in mind that the most efficient way to remove the data from the table may not be to just run a DELETE statement on the data itself. If you are fortunate enough to be working with a properly partitioned table, it may be as simple as simply sliding the partition completely out of the sliding window.

If you are not working with partitioned data, the options are a bit cloudier, but one that can be sneakily efficient in many scenarios is to create a copy of the table, transfer only the necessary data to the new copy, delete the old table, and rename the copy to the old name. It is good to be cognizant of foreign key relationships in this scenario, but this type of removal can be superior to the generally poorly performing DELETE FROM SomeTable WHERE data < SomeDate. Another solution that can perform very well in certain scenarios would be to add something like the ability to do soft deletes (i.e., an isDeleted flag) to allow you to do batch deletes and prevent massive table locks and things of that nature.

While this section has so far focused on drilling into what data to remove unneeded data and some recommendations for how to do that, we should not leave this section without returning to the topic of regulatory requirements. While your technical personnel may be generally aware of things like Sarbanes-Oxley or GDPR requirements, they may not be aware of every regulatory and contractual requirement governing the data we are trying to cull prior to migration. This may be especially true of consultants who are not accustomed to working with the multiple layers of compliance personnel that companies in heavily regulated industries tend to have. Whether the project is all consultants, all internal personnel, or a blend of the two, asking questions about contracts and regulations governing this data is critical to the eventual success of the project.

Note   Be aware of, and ask questions about, regulatory and contractual requirements that govern the data to be migrated.

You do not want to successfully execute a migration (especially if you've removed data in the process) only to find out that the team has run afoul of some requirement that has knocked your company out of compliance with something they very much need to comply with. Our goal is to get all of this done quietly with minimal impact – tripping up an auditor has a tendency to get very loud and very impactful quickly!

Understand Usage Patterns

So far, this chapter has focused on a generally well-organized project and complementary, fairly compliant project team. As you might guess, this is not always the case. So what should the technical team do if the business users are unwilling to help or just do not know the data at a deep enough level to know what can be removed and what should not be removed?

While this is certainly not preferable, there are some tools in the toolbox that our tech team can take advantage of in this situation. If we are fortunate enough to be working with a single database per application, it may be fairly straightforward to examine logs (SQL Server, application, etc.) for the most recent logins and user actions. If an application or its database hasn't been logged into in months or years, that is likely a good indication that this database may be a good candidate for migration or, perhaps, complete archival and removal. It at least gives the technical team the ability to go back to the business owners, point out that the application hasn't been used in years, and nicely (or not so nicely) ask for permission to remove the database.

If we are not fortunate enough to be dealing with a discrete scenario like that, there are still some tools available to us to see whether or not the databases (or parts of it) are being used with any regularity. SQL Server contains two Dynamic Management Views (DMVs) that are very useful to our DBAs as they perform this analysis. *sys.dm_db_index_usage_stats* and *sys.dm_db_index_operational_stats* allow a DBA to see how often an index is used. Unless you have a database full of heap tables (and it is my sincere hope that this is not the case!), your DBAs will be able to discern query patterns and data access patterns from a review of what indexes are being used and how often.

While these DMVs can be used in similar ways, there is an important distinction between them that is worth highlighting here. Dm_db_index_usage_stats records how many times the query optimizer includes and uses an index in a query plan. As one might guess, each time a plan is executed, the information within the DMV is updated. If a plan is executed but an index included in the plan is somehow not used, it still counts as a usage of the index. Dm_db_index_operational_stats records how many times the engine itself executes a plan operator that touches that index. For this reason, our DBAs may find the operational stats DMV more useful to figure out which indexes (and thus which tables) are being used, but to perform a complete analysis, a blending of the information from both DMVs is recommended.

There are many ways you can use system tables and T-SQL queries to determine which tables and indexes are disused. We can find tables and indexes that are unused within a given database, say, the **WideWorldImporters** database, by using the DMV **sys.dm_db_index_usage_stats**. First, know that DMVs do not carry forward data past a reboot. Consequently, we are unlikely to know about unneeded tables from years in the past, since most Windows servers are rebooted at least once or twice per year. However, we can also use code to determine how long it's been since our last reboot, as shown in the following:

```
-- How long since the server was last rebooted?
SELECT sqlserver_start_time
FROM sys.dm_os_sys_info;

-- Get each user table and all of the table's indexes
-- that haven't been used by using a NOT EXISTS against
-- sys.dm_db_index_usage_stats. INDEX_ID = 1, clustered
-- index. 2 or greater, non-clustered indexes.
```

```
SELECT DB_NAME() AS DATABASENAME,
       SCHEMA_NAME(A.SCHEMA_id) AS SCHEMANAME,
       OBJECT_NAME(B.OBJECT_ID) AS TABLENAME,
       B.NAME AS INDEXNAME,
       B.INDEX_ID
FROM   SYS.OBJECTS A
       INNER JOIN SYS.INDEXES B ON A.OBJECT_ID = B.OBJECT_ID
WHERE  NOT EXISTS
           (SELECT *
           FROM   SYS.DM_DB_INDEX_USAGE_STATS C
           WHERE DATABASE_ID = DB_ID(DB_NAME())
             AND B.OBJECT_ID = C.OBJECT_ID
             AND B.INDEX_ID = C.INDEX_ID)
   AND  A.TYPE = 'U'
ORDER BY 1, 2, 3 ;
```

Using the preceding code, we can see that the on-premises SQL Server has been running continuously for more than a year, a pretty good margin for ascertaining whether the indexes of WideWorldImporters are used or not. The second query produces a result set as shown in Figure 6-1.

	DATABASENAME	SCHEMANAME	TABLENAME	INDEXNAME	INDEX_ID
1	WideWorldImporters	Application	Cities	PK_Application_Cities	1
2	WideWorldImporters	Application	Cities	FK_Application_Cities_StateProvinceID	2
3	WideWorldImporters	Application	Cities_Archive	ix_Cities_Archive	1
4	WideWorldImporters	Application	Countries	PK_Application_Countries	1
5	WideWorldImporters	Application	Countries	UQ_Application_Countries_FormalName	2
6	WideWorldImporters	Application	Countries	UQ_Application_Countries_CountryName	3

Figure 6-1. *Unused clustered and nonclustered indexes from WideWorldImporters database*

As mentioned earlier, these tables and indexes might not need to migrate to your Azure landing zone. Consult with the business analysts and other members of your team to determine whether those tables should be archived or migrated.

Considerations on Database Design

The database migration tools provided by Microsoft, such as the SQL Server Health Check mentioned in Chapter 4, also assess your on-premises SQL Server databases for any gross violation of time-tested best practices. However, a quick review of those best practices is included here in case you haven't already acquired this knowledge. While there are a very large number of database design best practices for SQL Server and Azure SQL, we will delineate the most important here.

The most important best practices when defining keys and indexes include the following:

- Each table should have a primary key.

- Each table should have declared foreign keys on columns used in JOIN clauses.

- Each foreign key should have a nonclustered index.

- Columns frequently used in WHERE clauses should have a nonclustered index.

- Each table should have a clustered index, most likely the primary key (although not always).

- Each clustered key is recommended to use the BIGINT data type.

- Avoid using GUID *uniqueidentifier* for clustered indexes (use BIGINT).

- Avoid concatenated keys for clustered indexes (use BIGINT).

- Do not duplicate indexes.

When defining columns within a table that don't require a special data type:

- Avoid deprecated data types like SQL_VARIANT, IMAGE, and TEXT.

- Avoid variable length data types of just 1 or 2 in length, for example, VARCHAR, NVARCHAR, or VARBINARY.

- Avoid foreign keys on a column that are untrusted or disabled.

- Ensure that columns of a given name that appear in multiple tables and programmable modules are of the same data type; for example, a column called email is VARCHAR(128) everywhere it is referenced.

- Fixed-length columns that are grossly oversized.

We haven't taken the time to fully discuss the rationale behind each of these best practices. But these are all well-established best practices based on long years of experience in the industry. In fact, when you use the database migration tools offered by Microsoft, the tools will test for these best practices and many more. And one of the best features of these tools, like the SQL Server Health Check or the Azure Database Migration Service or the Database Migration Service (described shortly), is that they explain the rationale for these and many other best practices involved in a cloud migration with links to further reading that can fully educate you on a wide variety of best practices.

Migrate Meaningful Code

When you are discussing any portion of a data migration, it can be very easy to consider all the "physical" assets you are migrating. A lot of times the data in the database(s) falls into that category, and that is why this chapter begins with discussing migrating meaningful data. Our minds almost always think of it first. That said, it can be easy to forget the code that lives in stored procedures, functions, etc., within that database.

Even if we are not forgetting those objects as we plan and design a migration, we may not be giving their analysis the attention that it deserves. I have seen many successful "physical" migrations derailed when it is discovered that some piece of code is unsupported on a newer version of the database or behaves/performs far differently than expected.

Sometimes, that kind of an oversight can severely impact the project because it is quite likely to force a quick fix or, worst case, rolling back the migration completely until further research, testing, and development can correct the problem. Having participated in projects where this happened, I can confirm that it is no fun at all. It is an uncomfortable conversation to have when an oversight like this negatively impacts a project, especially when there are many tools at our disposal to prevent it. The remainder of this chapter will walk through the wide variety of things we can use to ensure no surprises from the code we are migrating alongside our data.

Data Migration Assistant

One of the easiest (and most important) things we can do prior to a migration is evaluate the database's suitability for an upgrade in version and/or a change in the platform (SQL Server, Azure SQL, etc.) on which it runs. It is a proactive step toward preventing an embarrassing issue with the migrated database.

To this end, Microsoft created the Data Migration Assistant (DMA). Look for more information and download links at `https://docs.microsoft.com/en-us/sql/dma/`. It enables us to assess a SQL Server (or AWS RDS for SQL Server) database for migration suitability to SQL Server (a different version), SQL Server on Azure Virtual Machines, Azure SQL Database Managed Instance, or Azure SQL Database and its various service tiers.

Before we dive into DMA itself, a note of caution before we proceed. While the specific permissions have changed somewhat over the years, the various evolutions of the DMA have often required several elevated permissions in Azure to correctly execute everything that DMA wants to run. It is good to be mindful of that and be careful granting those permissions broadly. If you are fortunate enough to work with a security team, it would be good to review this with them to ensure that an appropriate security posture is used for the account(s) running DMA. Now, let's carry on with running DMA.

The first thing you will notice when you create a new project is that there are two project types listed: Assessment and Migration. You will note that many of the DMA resources found within the data platform community in presentations, blogs, etc., focus on the assessment project. That's with good reason, as that is arguably the first step you need to execute when considering a database for migration. Indeed, let us start there as well before we discuss the migration abilities of DMA.

A DMA assessment project analyzes the database (and the data and objects that it contains) in two different ways: feature parity between the source and target versions of SQL Server and compatibility issues that may exist between the source and target versions (and platforms) of SQL Server and Azure SQL. The feature parity analysis performed will identify issues that will block your migration but, helpfully, also recommend workarounds and/or mitigation work that can be performed to remove that obstacle. In a similar vein, it will identify both partially supported and completely unsupported features and provide workarounds where appropriate. Performing an assessment within DMA is a crucial step to perform in order to know what additional work will need to be performed on a database before it is a proper migration candidate.

Hand in hand with the feature parity analysis is the compatibility analysis that it performs between your source and target. This is broken down initially by compatibility level and then into four distinct categories under each compatibility level. This can be helpful because it is quite possible to have a scenario where you need to migrate databases because a data center is closing, on-premises hardware is being retired, etc., but the database is simply not ready for the latest SQL Server/Azure SQL compatibility level. DMA's interface allows you to find a happy medium between migrating to a newer version or newer platform while also supporting legacy code.

The compatibility analysis appears in DMA with each compatibility level on its own tab when the compatibility issues radio button is selected. Simply click each tab to see the four categories the analysis is broken down into. Those four categories are breaking changes, behavior changes, deprecated features, and information issues.

Breaking changes are clearly defined as something that simply will not work if the database is migrated to that platform and that compatibility level. As with feature parity, workarounds and mitigations are suggested where possible. *Behavior changes* are not things that will break if migrated to the target, but their performance may change in ways that you should be, at a minimum, aware of. *Deprecated features* are exactly that – features detected in your database that have been deprecated at later compatibility levels. Finally, *information issues* are issues you need to be aware of and are typically recommendations to modernize your choice of data types or something similar, much like the best practice section earlier in this chapter. They do not require immediate action but should be slotted into the development queue as appropriate.

A migration project within DMA eventually does exactly what it sounds like – it migrates a database from your selected source to your selected target. It will generally suggest performing an assessment before doing so, and as I hope the previous paragraphs made clear, this is highly recommended.

It is a bit beyond the scope of this book to offer a step-by-step guide on how to perform each type of migration. Further, this tool has undergone significant evolution over the years, and that is unlikely to change. That said, you do have two main migration paths if you elect to use DMA to perform a migration. One path is to perform the migration within DMA itself (although, at press time, not every Azure SQL migration target option was supported for migration within DMA). The second path is to upload the migration project from within DMA to the Azure Database Migration Service (DMS). DMS is a superset of DMA and will provide the assessment features of DMA, as well as additional capabilities such as moving the data from your source database to your Azure landing zone. Read more about DMS at `https://docs.microsoft.com/en-us/azure/dms/`.

Similar to DMA, the DMS has undergone significant evolution over the years, and there is no sign of this stopping. Any sort of step-by-step guide to walking through DMS, also known as Azure Migrate, may be outdated by publishing time. Like we said, it changes quickly and frequently. Microsoft helpfully offers excellent step-by-step documentation for just this purpose. That said, Azure Migrate is increasingly popular as a "one-stop shop" for migrating databases, database servers, and their associated app and web servers. It is a very common path to use DMA for the feature parity and compatibility assessments and then upload that project file into Azure Migrate to execute the migration itself.

One additional thing to keep in mind with this "one–stop shop" approach is the differing SLAs of the various services. Given that these can and will change, we are not publishing a comprehensive list here. That said, it is important to research the SLAs of your various components (databases, VMs, services, etc.) to have a comprehensive picture of the SLA of your migrated environment (and whether or not Microsoft financially backs the SLA at all levels of your solution).

Finding Issues in Your Database Pre-migration

Since monitoring has already been covered in Chapter 4, this section serves as suggestions for what types of antipatterns you should be using your monitoring tools to identify and help remediate prior to migration.

First off, it is always easiest to identify the low-hanging fruit, and that is a good place to start these recommendations as well. This is also an area that our monitoring tools, regardless of what we are using, are typically very helpful in as well. We strongly recommend that you find long-running queries, resource-intensive batch jobs and transactions, etc., because things like that tend to be resource intensive (which costs us money in the cloud). They also tend to be easier to fix as there is a lot of room for improvement! Most monitoring tools excel in identifying these parts of your workload, and it is where we should begin.

Secondly, and unfortunately, we cannot stick with the easier parts of this forever. What we can do is next focus on things that may not be difficult to find but should yield big wins in performance improvement. *Implicit conversions* can be a silent killer for SQL Server database performance and therefore can offer a significant improvement if removed.

If implicit conversions is not a phrase with which you're familiar, a short layman's description is when a query asks SQL Server to compare two data types that are not the same. SQL Server will do this behind the scenes, dictated by the rules of data type precedence and compatibility, but performance can suffer significantly. When this happens, a flag is switched within the query plan itself, so any monitoring tool that is storing (and hopefully analyzing) query plans will make this type of research fairly straightforward. From there, a modification of the table or the query (or both) to compare equivalent data types should yield a big performance win that the team worked harder to fix than to find.

Next, as we work our way through the rest of the recommendations, they admittedly become more difficult to implement. The next recommendation goes hand in hand with the previous one, though, because a review of the query plans via a monitoring tool should reveal to us which servers have massive key lookup counts and what queries are triggering those *key lookups*.

If key lookups are an unfamiliar concept to you, they can be most simply described as when SQL Server needs to use a nonclustered index to retrieve the information the query has requested but there is no nonclustered index that covers every column requested. It then has to involve the clustered index in the retrieval of data, and this operation (and the operators that accompany it) is not kind to database performance. When these are identified, they can be remediated by (a) changing the query to better refine the request to use indexes that already exist or (b) adding a covering index for the query triggering the key lookup.

The DMV *sys.dm_db_index_usage_stats* can help us see how many full table scans and how many key lookups happen over time. The following example query shows how to retrieve this information within the scope of the database where it is executed:

```
SELECT DISTINCT
      db_name(database_id) AS 'Database',
      object_name(object_id) AS 'Table',
      index_id AS 'Index ID',
      user_scans AS 'Full Table Scans',
      user_lookups AS 'Key Lookups'
FROM sys.dm_db_index_usage_stats
WHERE (user_scans > 0 OR user_lookups > 0)
  AND database_id = db_id();
```

Figure 6-2 shows the result set when this query is run in the demo database WideWorldImporters:

	Database	Table	Index ID	Full Table Scans	Key Lookups
1	WideWorldImporters	BuyingGroups	1	2	0
2	WideWorldImporters	BuyingGroups	2	3	0
3	WideWorldImporters	CustomerCategories	2	5	0
4	WideWorldImporters	Customers	1	2	0
5	WideWorldImporters	DeliveryMethods	1	4	0
6	WideWorldImporters	DeliveryMethods	2	5	0
7	WideWorldImporters	People	4	10	0
0	WideWorldImporters	StateProvinces	1	2	0

Figure 6-2. *Indexes experiencing full table scans and key lookups*

In Figure 6-2, we can see that the table BuyingGroups has two indexes by looking at the first two lines of the result set. In line 1, the value where IndexID = 1 represents the clustered index on the table, while IndexID = 2 is an additional index, which is a nonclustered index. Note that when using this DMV, heaps are always represented by a value of 0, clustered indexes are always represented by a value of 1, and values of 2 or more indicate nonclustered indexes.

Another place where monitoring tools can greatly cut down research time as we work to improve workload performance is in identifying *deadlock* issues for us. While SQL Server does record and retain these and they can be accessed via extended events (Xevents), monitoring tools typically offer a friendlier interface than SQL Server Management Studio (SSMS) to review the deadlock and the queries involved. Deadlocks are not silent performance killers like some of the previously mentioned antipatterns, but they certainly cause significant periodic disruption in your system that affects your broader workload as queries queue up waiting for the deadlock opponents to sort out their conflict.

There are many ways to identify deadlocks, such as extended events mentioned in the preceding paragraph. However, the following query retrieves the number of occurrences of a few important usage counters for the SQL Server instance where you run the query:

```
SELECT (rtrim([object_name]) + n':' + rtrim([counter_name]) +
    n':' + rtrim([instance_name])) AS 'PerfMon Ctr',
    [cntr_value] AS 'Value Since Last Restart'
```

```
FROM sys.dm_os_performance_counters
WHERE [counter_name] IN
      (n'Forwarded Records/sec',
       n'Full Scans/sec')
  OR ([counter_name] = n'Number of Deadlocks/sec'
      AND [instance_name] = '_Total')
ORDER BY [object_name] + n':' + [counter_name] + n':' +
      [instance_name];
GO
```

The preceding query retrieves the accumulated occurrences, since the last server restart, for forwarded records/sec (an important indicator of problems with heap tables), full table scans/sec (neither good nor bad, but might point to high I/O queries), and deadlocks/sec from the DMV *sys.dm_os_performance_counters*. Like other DMVs, these values accumulate since the last server restart. So you will need to capture a delta of the values by polling multiple times throughout the day. Figure 6-3 shows the result set for this query.

	PerfMon Ctr	Value Since Last Restart
1	SQLServer:Access Methods:Forwarded Records/sec:	0
2	SQLServer:Access Methods:Full Scans/sec:	136501389
3	SQLServer:Locks:Number of Deadlocks/sec:_Total	0

Figure 6-3. *DMV values for workload-relevant performance counters*

While Figure 6-3 shows the results of the query retrieved from the DMV, you see that the data retrieved is from a single point in time for the entire instance of SQL Server. If you want to monitor the number of deadlocks your application experiences in real time, we recommend that you create an extended event for deadlocks. Details and instructions are available at https://docs.microsoft.com/en-us/sql/relational-databases/extended-events/quick-start-extended-events-in-sql-server.

Next, using a monitoring tool to identify warnings within query plans about *missing indexes* or *missing statistics* can be a quick path to some nice performance improvements with minimal work by the team. Depending on the tool, these may be identified within a graphical display of the query plan and/or a separate part of the tool that focuses on missing indexes, outdated or nonexistent statistics, and similar. In many cases, but not all, remedying issues like this is as simple as creating the missing index, creating the recommended statistic, or tweaking your settings around how often statistics are updated so they stay current.

With regard to indexes, however, just because SQL Server suggests does not mean it will help a great deal and it might cause significant issues to other parts of your workload. It is imperative that the team reviews these index recommendations and tests them extensively before broadly applying them.

The final performance killer to inspect via your monitoring tool of choice is to look for query plans with multiple SORT operators, multiple HASH JOIN operators, or multiple DISTINCT operators within the plan. While refining code that contains some (or all) of these is not a simple process, assessing and refining your code for things like this is helpful for the migration and likely the best time to perform this kind of work.

Sorting through your workload for things as "obscure" as operators like this is quite often work that is consistently deprioritized for other things. Sometimes, that deprioritization never stops, and this work is never completed. Using the "must is a mighty master" of an impending migration to perform this kind of work can work to the advantage of the DBA team and, indeed, every database user when performance gains are realized in the database.

Baseline Testing

While many companies may have a commercial or home-built load testing suite that they prefer to use, that will likely require development and QA resources that may or may not be available to the migration project team. Given that possibility, let us briefly discuss a couple of easily available tools that will help our DBAs and data people perform this type of testing. If we are able to supplement this with full-blown load testing from other parts of the company, even better!

The first tool we'll review is the Database Experimentation Assistant (DEA) from Microsoft. While this tool has seen varying degrees of investment and promotion, it can be very useful to a team who doesn't otherwise have access to resources to test their workload. Microsoft describes DEA as an A/B testing tool, which basically means that you can evaluate the performance of your workload in source and target without performing a true migration. Figure 6-4 shows an example of capturing a new workload trace using DEA.

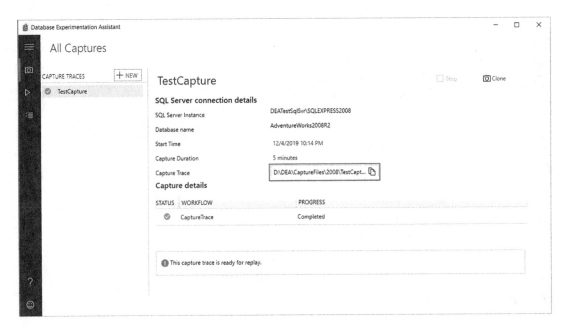

Figure 6-4. *DEA new workload capture*

How does it do this? By allowing you to upload Profiler traces or, most preferably, extended event files where it can then perform statistical analysis against those workload traces on both the source and target instances. You can then review data visualizations depicting the comparison to see if you're maintained performance or (hopefully) improved it by migrating to your choice target. If Azure SQL Database, Azure SQL Managed Instance, or SQL Server on Linux is in your chosen migration targets, this tool can help you evaluate how your workload will perform in its new home, as shown in Figure 6-5.

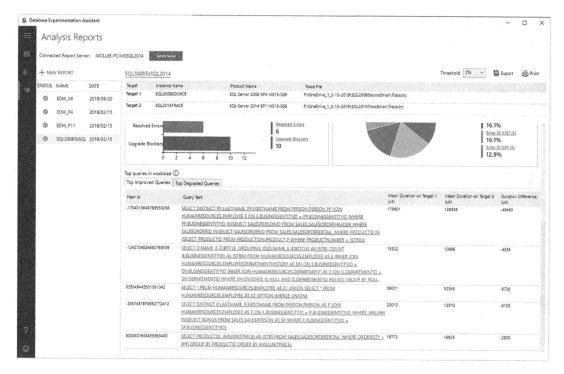

Figure 6-5. *DEA sample workload analysis report*

Figure 6-5 shows the sort of output you can expect from DEA. The report shows us the differences between two workload traces, with aggregate errors and bloggers shown at the top and tabs showing queries that improved performance between the two workload replays and, on a separate tab, queries whose performance degraded between the first and second workload replays. Examine the report carefully to ensure that you're not going to experience significantly worse performance on your target Azure landing zone.

If you would prefer a community-based tool to a vendor-based tool, WorkloadTools (found here: `https://github.com/spaghettidba/WorkloadTools`) is a suite of tools that allows you to perform detailed source and target workload comparisons. It is quite customizable and, due to being available on GitHub and MIT licensed, available for contributions from the public as well. It also uses Power BI for its data visualizations, which some people may find more appealing than other tools available to them. Thorough licensing and readme information is available in the GitHub repo, but we would be remiss in this book if we did not include information about WorkloadTools due to its flexibility and power.

That said, there is certainly nothing wrong with vendor tools, so let us close this section highlighting two of those (and a reminder that this is an ever-evolving market space). New Relic provides a fairly comprehensive view of your overall environment. As you might imagine with a comprehensive tool, you may not find every single monitoring data point at every level to your satisfaction, but there is real value in getting an overall picture of (and baseline for) your entire environment. Along with providing you fairly complete baseline information, the overall picture can help more accurately guide your troubleshooting when some or all of your deployed resources experience issues.

Finally, Azure Application Insights (commonly known as App Insights) can provide general (and in some cases, very specific) monitoring information across many of your deployed Azure resources. While App Insights is commonly understood to provide performance management and monitoring for deployed web applications, its overall incorporation into Azure Monitor is certainly worth a look as you review what your monitoring solution(s) should be as you migrate to the cloud.

As you may guess from the content of this section, there are many ways to baseline and monitor your migrated assets, and there is no one right answer to what the best solution is for everyone. In the classic tradition of many data-related questions, the answer to "Which monitoring solution is the best?" is definitely "it depends."

Remediation

Remediation of poorly performing code could be its very own book, so this section will just lightly touch on recommendations for how to approach this. You may notice that in this chapter, we have taken a very DBA/data team-specific approach to tackling identifying and remediating poorly performing parts of your workload. That is because in many migration projects, proper development resources may not be available. Data personnel may be left on their own to perform this work.

While the actual machinations of this code remediation will vary company by company depending on the size of the data team, their skill sets, the development methodology being used, etc., there is one key point to take away from this. Simply identifying these performance issues and telling just your DBAs to go fix it will likely not correct every issue identified during the workload performance review process. Project leadership must work to allow space in the development process for this work to be executed and tested. While everything discussed in this chapter is contained within the database, the ultimate success of the remediation will rely on development resources being available to, and perhaps integrated with, the data personnel tasked with issue identification and remediation.

Put more simply than that, just because the performance issue resides in the databases, it is not a lock that the DBAs can fix it without help. The path to a successful migration relies on the data team having what they need to correct the highest priority issues before you get your data to the cloud and overspend on running bad queries and returning bad data.

Summary

Careful consideration needs to be given on a migration to be sure that the data and code that are getting moved are actually needed. Migrating a database to the cloud provides us with a great opportunity to review and remediate any outstanding technical debt, poorly performing code, compatibility issues, or unnecessary data that could cost us when migrating. In a typical on-premises environment, hardware is provisioned to last for years and account for growth, so we are usually fine letting things slide for a time with the intention of coming back and fixing it later. With the cloud, you are paying for what you consume when you consume it. Put another way, postponing improvement work for months or years is going to cost you.

In this chapter, we first looked at what it means to move meaningful data. If your database is large, you should always evaluate the data that is stored in it to see if it should be included in your database migration. Oftentimes, you may find data no longer useful or leftover remnants of an application feature that never got fully used that may not make sense to move. The first thing you will want to consider here is the business requirements for the data. As a technical resource, you may need to engage with other members of your organization in making some of these decisions to make sure that you are meeting business and compliance needs. If you are unsure, then reach out and have conversations to understand this better. The next thing you will want to look at is the usage patterns of the database to better understand what is being used and what is not using methods we described in this chapter.

In the latter part of the chapter, we looked at what it means to migrate meaningful code. This means that we make sure our code is supported, compatible, and performative with the service we are migrating to. We looked at how the Data Migration Assistant plays an important role in helping us understand compatibility, how our monitoring tools and scripts can help us discover and remediate bad performing code, and how baseline testing using tools like the Database Experimentation Assistant can uncover unexpected differences. The remediation effort to correct this can take some time, but it will be well worth it in making sure you have the best experience and, ultimately, a successful migration.

## CHAPTER 7

Team Success Factors

Technical knowledge and performing due diligence play important roles in any cloud migration, but there are also several other factors that are equally important. Factors that impact the flow of a team can make or break any project. In this chapter, we are going to look through several team success factors that are important to consider throughout your project, many of which are also not exclusive to just cloud migrations but can be applied to any large IT project you are working on.

Before we can define what makes the team successful, we need to determine what makes a cloud data migration successful. At its simplest basic form, it is to successfully complete your migration. Or more specifically, it's taking the systems you designated for migration and operationalizing them in the cloud. From there, you can spin that off into other more detailed success criteria such as completing it within a specific timeline or keeping costs under budget, but the ultimate goal is getting it there successfully. For your team, there are several factors that impact productivity and morale of the team that can positively or negatively impact your migration outcomes. We will want to keep these factors in mind as we go through the chapter.

Business Value

One of the first steps in taking an idea and turning it into a project is determining its business value. There are two reasons why this is important. First, the business value is often weighed against the time and cost by leadership within your organization to determine if it's something worth pursuing. And second, it establishes the key goals and focal points for what you are trying to accomplish. If there is little or no business value in what you are doing, then you should question if it's something you should be doing at all.

Note If there is no business value in what you are doing, then you should question if it's something you should be doing at all.

© Kevin Kline, Denis McDowell, Dustin Dorsey, Matt Gordon 2022
K. Kline et al., *Pro Database Migration to Azure*, https://doi.org/10.1007/978-1-4842-8230-4_7

Before any fiscally responsible organization agrees to take on a project that has a budget tied to it, they are going to require some level of business justification. This means helping them understand the value to the business this project brings before determining if the project is approved. And usually the higher the investment needed, the more interested they are going to be in understanding the value. This is true for any project, but especially true for larger direction-changing approaches, such as a cloud migration. Unless your leadership is already sold on the idea, then the first step for any team is working to win their support. This starts with being able to highlight the business value.

If your leadership is already convinced this is the path forward, then they probably already have a solid understanding of the benefits and value to the organization, which makes your job a little easier. (Be careful when leadership is convinced to undertake a cloud migration for the wrong reasons, as we discussed in the cautionary tales of Chapter 1). However, the rest of us have a little more work. The benefits of the cloud are widely available and shared, so it's often just taking those items and using them to tell a story for your organization. Several of these benefits are shared throughout this book, especially in Chapter 1. But let's take a look at some of the most common benefits the cloud offers us:

- Faster time to deployment

- Ability to pause/resume and scale resources on demand

- Only pay for what you consume

- Potential for lower costs

- Increased accessibility

- Simplified administration

- Increased security

- Improved High Availability and Disaster Recovery

If the business value is not immediately clear, then look through this list and you can surely begin to craft your story. For example, let's assume you are migrating a database that runs a critical CPU-intensive batch process every morning during a 1-hour window and the rest of the time usage is low. In an on-premises scenario, you would likely allocate enough resources to support the 1-hour window with no real need for it outside of that. In the cloud, we could use the benefits "Ability to pause/resume and scale resources on-demand" and "Only pay for what you consume" to show business value to our organization. In the cloud, we could pause/resume/scale during periods of higher resources versus times lower resources are needed and only pay for the higher consumption when needed.

While sharing and stating the business value is important to getting the project off the ground, it's also important that the team always keep this in mind throughout the project. Details on projects can often change after conception, but the goals should generally stay the same. Remembering the value of why you started doing what you did can help you stay on track to your commitment to your organization.

Business Sponsorship

Another important role of a team's success is to identify and obtain an *executive sponsor*. An executive sponsor is someone within your organization that has direct influence over organizational decisions that are made. Generally speaking, this would be someone in upper management or part of the executive team. But depending on the structure and size of the company, this may not always be the case. With some big projects, the executive sponsor can naturally occur because the direction comes from the top. However, when technical or nonbusiness groups try to promote a project, then this may not always be the case. If you do not have a natural fit for this, then you may need to reach out and engage with leadership to find someone. For technical projects such as a cloud migration, a CIO or a CTO are great candidates.

The primary role of the executive sponsor is to help bridge the gaps between the teams performing the work and the stakeholders or business, often to unblock stalled processes or break deadlocks over direction. Technical and business people often view goals and success factors differently, so having someone with influence that can speak to both sides is very helpful. Let's take a look at some of the specific benefits that an executive sponsor provides.

Note   The primary role of the executive sponsor is to help bridge the gaps between the teams performing the work and the stakeholders or business.

They **assist with obtaining resources**. With any large projects (especially ones that are new to the team), there is always going to be some bit of unknown. As much effort as you put in estimating what you need, there is always the chance that something comes up. For instance, you realize you need additional staff, training for a particular skill, a new tool, or anything else that may impact the success of the project and then the executive sponsor becomes your go-to in these types of situations. You can work with them to communicate the need, and they have either the power to approve or the influence needed to address the issue. This does not negate the need for good planning ahead prior to the start of the project or mean that you think of your sponsor as an ATM, but they are there to help when needs arise.

They **make sure that the project stays aligned with the overall strategy**. As the project evolves and changes are made, the sponsor can help make sure your vision stays aligned with the vision of the business. They understand what success looks like from both a business and a technical sense and can make sure both stay aligned. They also keep other leadership informed of the status of the project.

They **serve as an escalation point**. Throughout a project, you may hit obstacles that you are unable to handle yourself and need someone with a higher rank to step in. This can entail a myriad of things but can be especially important when you have people or resources involved. For example, you have a dependency on another team that goes through a different reporting structure and has different priorities. The sponsor should be able to address this by having the ability to speak directly to both teams or engaging the right person that can.

Being a sponsor is generally not something that is a full-time role, and these folks have other demands on their time. The business and financial impact may play a big role in their level of involvement and how they perceive it. Regardless, if a project warrants an executive sponsor, then it warrants their attention. It's generally in their best interest to stay engaged, because they will often be judged based on the success or failure of the project by their peers and leadership. Poor results could negatively impact their careers, whereas positive results could enhance it. As a technical lead on a project, you might have to work to keep them engaged if they are not doing so naturally so you have the benefits listed just before this.

To keep your sponsor involved, there are a few things you can do at the onset of the project. First, set clear expectations for how you expect to utilize them so they know what to expect. Second, establish a communication plan that will continue throughout the end of the project. This may mean that you set up a meeting cadence with them to discuss the status of the project and if there are any roadblocks you need assistance with. Communicating via other methods such as email, text, and chat in between meetings can be effective. However, we still suggest a regular cadence to speak directly. This avoids things getting overlooked or lost in translation.

Support of the Team

To build momentum and keep the project moving forward, it helps to have the support of the teams involved who need to complete the work. This creates cohesion and excitement and can enhance the morale and productivity of those involved. When you have key members that are disgruntled or disinterested, then it can negatively impact forward progress. For a cloud migration, you will most likely need to work and depend on other team members for part of the work, and this goes a lot smoother when you have buy-in from all those involved.

Early Involvement

One of the most frustrating things that can happen with technical teams is when someone else makes a technical decision without ever getting input from the people who know it best. If you are a technical subject matter expert (SME) for your organization, you expect that leaders want your input ahead of making any critical decisions related to it. It makes too much sense not to.

However, for anyone that works in IT, you know this is not always the case. Throughout our careers, we have had leaders commit to projects with significant spend and tight timelines in our area of expertise without our input more times than we can count. Usually, we are not even made aware of it until it's time to implement a strategy, at which point we discover several issues that would have been better addressed at the onset. What we are left with is a disgruntled and frustrated team who is scrambling to shift priorities to meet unrealistic deadlines. No one likes to be in this type of situation.

When taking on any big project, the workflow progresses much better when you establish a broad consensus, involve leadership and teammates from the project's inception, and strive to actively involve all members of the team. In our experience, team members that are involved in the planning and architecture stages of a project are a lot more invested in the overall success of it. While you may not be able to change the way your leadership approaches projects, you can change the way you do.

If you are pioneering a project such as an Azure migration (or playing the role of project manager), then try to involve as much of the team as you can in some of the early conversations. This helps build a sense of ownership and excitement that may be harder to achieve down the road. Try to think of a time when you were included early on in a project versus a time you were brought in later. Do you see a difference in how you viewed them?

Note   Team members that are involved in the early planning and architecture stages of a project are much more invested in its overall success.

Having a talented and dedicated project manager can alleviate a lot of these concerns by making sure the right staff are involved at the right time. They do this by having the foresight to look at the big picture and stay ahead of any potential roadblocks due to someone pertinent being out of the loop. For large IT projects such as cloud migrations, we highly recommend having a dedicated project manager to keep track of and facilitate tasks on your critical path. However, we also understand that it is not always possible to employ a professional project manager for a variety of reasons. We frequently see IT projects where IT staff or leaders are expected to manage their own projects. And this is when problems arise.

A cloud migration project is an exciting project for most teams. So generally, it's easy to get the team excited about it. This is especially true if it's your organization's first venture into cloud technology. The cloud gives you access to cutting-edge technology with all of the latest features without the complications associated with long-standing on-premises applications and databases. Additionally, it is a great opportunity to grow your career by gaining experience with highly in-demand skills. Even though cloud technologies have been around for a while now, the overall workforce is still limited with experienced candidates. So it's a great way to help set yourself apart for future job searches. As a leader or as a member of the migration team, these achievements should be celebrated and promoted to build excitement.

Encourage Growth

I have worked with managers in the past who have intentionally limited their team's personal and professional growth out of fear of them leaving for another job. This meant limiting learning activities or avoiding the latest technology that could expose an employee to other career opportunities. From a leadership perspective, they spend months or even years trying to fill their role so they don't want to risk losing them. While the reasoning is understandable, this is a very selfish viewpoint that puts the needs of the leader over the employee – not a good leader.

If you are a manager and that is your mentality, then shame on you. If you are an employee that works for someone that has that mentality, then find another job. A good manager should always strive to provide the best opportunities they can for their staff, even when it may come at a cost to them. One good measuring stick of success for a manager is by viewing the success of their staff, both past and present. For example, is there a high number of promotions or are staff that have quit in higher positions elsewhere?

Note   A good measuring stick for determining the success of a manager is by viewing the success of those that worked for them.

We have lost several very talented employees and colleagues over the years to better opportunities, and, yes, it stings. It's tough to lose people, especially when they are talented and a pleasure to work with. But one thing we always do is celebrate and be happy for them. We take a lot of pride in knowing that we played at least some part in their journey to help achieve career success. Rather than looking at the negative side, we use it as a feather in our cap. When we begin backfilling that role or even other roles, remember to share the success stories of past employees as a selling point to show the type of growth you can have under your leadership. This could be the difference between getting the ideal candidate versus the candidate you settle with.

As a leader, be the person that encourages and inspires growth of those around you. Look for and provide opportunities to build their skills or offer ways to utilize newer technology. If you are reading this book, then you probably are engaged in or have some interest in cloud technology, which is an excellent growth opportunity for both you and your team.

While we personally have worked for some poor managers that limited our growth, we have also had some great ones. Our best managers have invested and believed in us when they had little reason to do so. They provided us training opportunities (including paid ones), encouraged us, and gave us challenging projects that pushed us past our limits. If it had not been for them, then we know we would not be where we are today.

Dealing with Negative Energy

Sometimes, when taking on a new project that introduces new technologies, you have people on the team who oppose it. There are usually few groups of people that fall into this category: those that just disagree with it, the old timers, and those that fear for their job. The first category is people who are either contrarian, in general, or who might have well-reasoned opinions to resist or undermine the migration project. The old timers are usually people who have spent a lot of their career working at your company doing things a certain way and who resist change. The third group is people who fear that their job may be in jeopardy as a result of the project. For example, your project will make their job obsolete. Failure to address these issues can impact morale and the overall success of the project. Let's go through some different ways to overcome these issues.

First, you need to **have a plan**. The people who oppose change or may be negatively impacted by it should not come as a surprise if you have any working relationship, meaning you should be able to easily identify these individuals very early in the process. While it's important to think through every member of the team's role within the new project, careful consideration should be given to those that may oppose you, warranted or not.

For the old timers, it may be just a task for getting them on board with the path forward or excluding them from the new project altogether. For those fearful of their jobs, your approach may come down to the legitimacy of the concerns or not. If there is no reason to have concern, then oftentimes that issue can be resolved with good communication. If their concerns are valid, then you will need to determine how you deal with that. Is the person subject to staff reduction or are there opportunities to move them into a new role? Regardless of the outcome, ignoring the issue will only create problems. Don't delay. The next several paragraphs will provide some tips on how to navigate these situations.

Always lead with "why." When people question a project being done in a particular way, it is because they don't understand the why. They fail to see the greater value of the project and that it strongly aligns the goals of their organization. They often look at the project with tunnel vision and focus on how it impacts them without focusing on the larger plan. Leaders should clearly articulate why a project is beneficial and the reasons behind it. It is helpful and even comforting to understand the larger context and the impact if we stay the course. A common mistake that we have seen is when leaders assume that their staff fully understand the big picture, when in fact they often do not.

Communicate, communicate, communicate. Most issues that arise from opposition can be cleared with being transparent and communicating frequently. If you leave people to their own thoughts, then they have a tendency to focus on the worst of a situation. After all, if no one is saying anything, then they must be hiding something they don't want you to know. Take time to communicate honestly with your teams often, even at the risk of overcommunicating. While they may get tired of hearing or seeing you, they will never question where you stand and how the project is progressing.

Get input from the team. This really goes along with communication, but reach out to team members who have an issue with a change. Strive for consensus at each major decision point. Get their input and address their issues directly. Acknowledge their feelings and be open to feedback. Sometimes, people can feel better simply by knowing their voice is heard. If your team is a large one, then seek out cheerleaders who are willing to carry your message forward. Many times, people who carry negative energy are already resistant to leadership but may be much more receptive to communications from peers and those they see as equals on their team.

Accepting the bumps with change. People generally enjoy being in comfortable and familiar situations and dislike being outside of their comfort zone. A lot of this is due to the fear of failure or making a mistake. Try to create a safe environment that insulates your team from these fears. Take steps to protect your team so that they are comfortable taking worthwhile risks. If people don't feel like they are able to make the occasional bad decision, they will not be willing to risk the good ones. We have all had to start somewhere, and at some point, we are where we are at with our careers because we stepped outside of or were forced out of our comfort zone. Remember that fact when you ask others to step outside of theirs.

And lastly, **provide people with the tools and resources they need to succeed with the change**. Make sure that you are asking people to step out of their comfort zone and take on a new challenge that you provide them with adequate time and resources to be able to succeed with it. This could include training, shadowing other teams, specific tools, or anything else that may play a part.

Promote Collaboration

In any IT environment, it is common to have different departments that consist of experts in specific areas. For example, you may have a networking, storage, database, virtualization, hardware, or security team to list a few that contain the corresponding subject matter experts. While you are all part of different teams and may have different reporting structures, you know that you need each other to be successful at your job. For instance, if we need a new SQL Server built, we are going to need involvement from these other teams. Whether it's provisioning a server and storage, assigning IP addresses, applying an image, or passing a security scan, you are likely relying on others for some part of this. Fostering good relationships and effectively collaborating with these teams can go a long way in your own personal success.

You may consider yourself a lone wolf and view migrating to the cloud as a way to remove ties to others. Eliminate the need from these other teams and just be able to handle it yourself going forward. While it's true that cloud providers do simplify the experience by making it easy to get the services you need, it does not mean it's fail proof. You may be able to do something, but it does not mean you are doing it right. Even if you don't have to, it's smart to engage with experts from other teams to involve them in what you are doing. This builds consensus, acknowledges their skill in a positive way, and helps reinforce the social bonds of your teams. When you read a blog or watch a few videos, recognize that short-term learning experience doesn't trump someone with years of experience. While the cloud can be viewed as easy, every choice you make has an impact. A simple mistake could result in poor performance, unexpected costs, or even a security breach.

A common theme you have seen throughout this chapter is the importance of communication, and it's no different here. Good collaboration starts with having good communication. When you are engaging with other teams or people who have not been part of the project since inception, start by having a conversation about the overall project before just sending them tasks. Share the business expectations and value the migration project brings to the organization so they have some level of context of what they are getting involved with. Chances are if you are engaging with a new team, then they are going to have other priorities they are working on so if they are disinterested or telling you they are too busy, then utilize the executive sponsor to help navigate.

Also when you are working with other teams, make sure you are working toward common goals managed from a single plan. You will want to make sure these are clear to everyone involved and reviewed frequently to avoid wanderers. This starts with not just verbalizing these things, but having the goals and project plan in writing and providing the team access to it. Next, set up a meeting cadence with the team to review progress and track against the overall project plan. The frequency can depend on a lot of things, so make your best judgment call on how often.

If you are still unsure, then weekly meetings are usually a good place to start and adjust as necessary. Beyond that, encourage the members of the team to freely engage with each other outside of the scheduled meeting times. If you have a good project manager, then these things are probably already being taken care of. But if you don't, then this does not become less important. Oftentimes, the project lead handles this responsibility since they are most in tune with the tasks that need to be done. But you could instead delegate to a member of the team to handle this.

If we sit and think about some of the best projects we have worked on, we can find several common themes between them. One of those is that the projects involved us working with people we really liked to work with. The people we worked with had great relationships and friendships which extended beyond our work. Humans desire relationships because they make us feel good. Because they increase our confidence and self-esteem and make us feel important and appreciated. So it's natural this would also contribute to our successes at work. Most of us will spend more time with our coworkers than we will with our own families. So good working relationships are advantageous to all.

Note   A powerful technique to build strong working relationships is to praise the good work of your teammates. A well-timed and sincere compliment can encourage other teammates to achieve their own successes and, in turn, to offer praise to others.

Let's look at some of the advantages that having good working relationships can have. First, it will **improve teamwork**. No surprise here, but people that get along well with each other are more likely to work well together. Next, it will improve **employee morale and productivity**. People are happier working with people they enjoy being around and those that boost their mood, and this often translates to them being more productive. The last one is increased **employee retention**. When employees have close relationships with colleagues, this has a huge impact on their likelihood of leaving or, at the very least, actively looking.

We feel more attached to the company because of the relationships we have built and do not want those to go away. This is a win-win for both the employer and employee. The employee gets to work in an environment they love, and the employer is able to retain more employees. As a leader, finding ways to strengthen these relationships can have a big impact on your team and projects.

The last thing we want to share in this section is that good collaboration is about leveraging the strengths of those around you to work toward a common goal. Every person has different strengths that can add to the success of a project. And likewise, every person has different weaknesses that could negatively impact a project. Utilizing your team in a way that allows people to focus on what they are good at will positively impact the success of your project.

Note   Good collaboration is about leveraging the strengths of those around you.

Good Communication

Communication is a common theme within this chapter and something you will see mentioned a lot to the point it may feel repetitive. This is because it is so important. Good communication for a project means that employees have all of the information they need to be able to do their jobs and make sure that nothing is missed or left to interpretation during the entire project. Failure to do this can lead to confusion and conflict across the team and impact the overall success of the team and project. A good way to think of this is that good communication maximizes success and minimizes risk.

Note   Good communication maximizes success and minimizes risk.

I am sure all of us have stories where we have seen this to be a problem. You get handed a task with limited information and you complete it only to find out you were missing key information and did it wrong. Because of this, you ended up wasting time that didn't need to be lost if you had just had all of the information up front. It's frustrating and can negatively impact your morale. While we may not always be able to change this for those higher up the ladder than us, we can make the choice to make sure we are not the source of this issue.

This also means making sure everyone has clear priorities and is working toward shared goals. For a project involving only one team, this is not so difficult; however, if you are working with multiple teams across departments, it can get challenging. This can be overcome by making sure the messages are clear and everyone agrees on them.

Another consideration is making sure you are creating a forum or environment where people feel free to openly share ideas and thoughts. You never know who may have a breakthrough idea sitting in their head that they are too scared to share for fear of rejection. Encourage team members to engage in conversation and add their inputs and make sure these forums are judgment-free. By repeatedly practicing this, we have seen some of my most timid employees become very engaged and offer up amazing ideas that have saved time and money.

In these forums, you may find that you don't always agree on things, but this can be a good thing. Debating ideas can be very healthy and educational for the team and often leads to a better outcome. Creating an environment where this is acceptable is something we encourage. If you never allow your ideas to be challenged, then you are probably doing something wrong.

Architectural decisions are a common space for debate and typically lead to improved outcomes because there usually are lots of differing opinions. Allowing each person to speak why a particular idea is better or worse can change the overall path of a project or at least provide context to the team for why their opinion leans one way or another. This is literally one of our favorite conversations to sit and have with our teams. The only issue you have to watch for with this is keeping everyone level headed and refraining from personal attacks. As a leader, you should be the mediator to make sure the conversation stays on track and professional.

Note   If your ideas are never challenged, then you are probably doing something wrong.

Communicating is not just about the words you speak, but it's also about the words you hear. A good communicator is also someone who is willing to listen. Even though you may be in a position to call the shots, take the time to listen and receive feedback from the team to understand their point of views or issues. This allows you to look for opportunities to be challenged and addressed in areas you may not have been aware of. It is also an opportunity to empathize and show respect and appreciation for those involved.

Training for the Team

If you are adopting a new technology for a project, it is important to consider whether it's something you should provide training for the team on. This is especially true if there are long-term plans for the technology to be used, such as a cloud migration. A lot of the technical skills you have used throughout your career still apply when you start to work with cloud technologies. However, it's not a seamless transition and will require new skills. Infrastructure is infrastructure, and a lot of the same rules apply on-premises versus the cloud, but they are handled differently. To succeed, your team will need to learn how to use it quickly or mistakes could happen.

For example, take someone who has years of SQL Server experience and is at the top of their profession. They have spent their career working with relational databases and know them extremely well. However, their only experience has been SQL Server. How would that person fare if you dropped them in an Oracle environment and asked them to manage it? Oracle is just another relational database, right?

Chances are that the person would be lost and figuring things out for a period of time. They would probably know enough to eventually figure it out. But it's a new system that has a different way of doing things. Even though the end product may be the same as a SQL Server Database, getting there is going to be different. Taking this person's already established background, providing them training such as Oracle class would make them more effective a lot faster. The cloud is similar to this. The end product of servers and databases is the same; just how you get there is going to be different than what you may be used to.

Too many organizations rely on their staff to struggle through figuring out a new technology and then are disappointed when deadlines are missed or mistakes are made. IT professionals are really smart people and can absolutely do this, but at what cost? IT professionals are usually one of the most expensive teams from a staffing perspective. Many companies would rather have them spend significant time learning something rather than investing in a solution. Providing staff training opportunities reduces the overall time spent on a project, which allows your well-paid employees time to focus on more valuable activities. The cost of training is often fractional of the cost of time saved with hiring more staff.

Whenever we have an interview with a new company, one of the questions we always ask is whether or not the company provides paid training opportunities for their staff. It is a great question that we encourage others to ask as well as it is helpful in determining how a company will invest in your future. Usually, the answer is only forthcoming if it's yes; otherwise, it can vary. The most common answer we are given is that the company would do so if it was needed, which can be true or simply be hiding the real answer. When you ask for specific recent examples, then you can quickly discover the truth of the situation.

There are a variety of reasons why companies don't offer training, and it's usually due to a lack of priority. Sure, you could have budget and time constraints. However, you will always make time for tasks that are a priority. When you plan for a year, plan to make this a priority and set aside budget dollars and time for your staff. They will be grateful for it. If you plan for a new project that introduces new technologies to the team, make sure that you are allocating funds for training. It is amazing to us how companies can spend millions on a project with little hesitation and yet lose their mind when you ask to spend a few thousand to properly equip them for the job.

Note   If you don't have time to do an important activity, it's because you have not made it a high enough priority.

It is common knowledge that training helps equip the team, especially when starting something new. It does this by providing people with confidence in the project and expectations around it. Let me share an example. Earlier in my career, I was asked to build a data warehouse (collection of data marts) for an organization I was working with. The company wanted to be able analyze several data metrics around patient data at our hospitals. At this stage in my career, most of my focus had been working with OLTP database workloads, which are quite different from a data warehouse. I knew SQL Server really well and had some experience with Analysis and Reporting Services so I knew I could figure it out with time.

I requested training and the company accepted. They sent me to a week-long course with a well-known group to learn dimensional data warehouse concepts. The training was one of the best I have ever received, and I was immediately able to apply what I had learned to my tasks. I was able to build and start getting meaningful data out of the data warehouse months ahead of schedule and did it more efficiently than if I hadn't received training. The company paid less than $5,000 and saved considerable amounts of my time, allowing me to do other tasks that added value to the project. Without question, if you asked the leadership if it was worth it, it would unequivocally be yes.

Training can also be a huge morale boost because it shows your staff that you are willing to invest in them and provide them every opportunity to succeed, which can lead to higher rates of employee retention. As with the training example mentioned previously, I had a deep appreciation in the company for providing me the opportunity. Not only did the training help me with the task at hand, but it also enabled me to excel in future opportunities in my career, making it a win-win for both of us. They got the end outcome sooner and gained additional skills as a result.

Training also helps prepare the staff for more responsibility and could be what is needed to take the next step in their career. Employees like to see opportunities for advancement within an organization and appreciate seeing pathways to get there. For example, if you have a midlevel engineer on your team, what does the path look like for them to get to a senior role? Probably safe to assume it involves greater responsibility and deeper technical skills and providing training opportunities is a great way to fill in those missing skills. Offering these opportunities and providing pathways is a great way to improve employee morale and give something for people to work hard toward.

Asking for Help

Taking the first steps in a cloud migration can be hard, and it's likely you will run across challenges along the way that can't be overcome without a struggle. In these cases, it's important to know when to ask for help to avoid getting stuck in the project. In several places throughout the book, we have shared information about situations, but it's important to highlight again here.

Be sure to engage with your cloud vendor early on as they will be able to help support your migration and provide opportunities for technical and/or financial assistance. Remember that they have a vested interest in your success. It's in their best interest to make sure you succeed. If you have issues that are undermining the success of your migration, then we strongly recommend speaking to them about your obstacles to see how they can help. Several of the Azure migrations we have completed would not have succeeded without assistance from Microsoft that we received along the way. This is especially true when adopting new services where you are inexperienced.

If you find yourself constantly engaging your cloud provider for assistance, then it may make sense to bring in a consultant or consulting group who has the expertise to get you started. With any cloud migration, one of the hardest things about starting is establishing a framework that you can build on. There are resources that can help you, but be prepared for some trial and error. If timing is important to you, then consider working with outside resources at least in the beginning.

Celebrate the Wins Along the Way

Depending on the scope of the work that needs to be done, be sure to document and celebrate the wins along the way. Large projects like a cloud migration may be something that takes many months to complete. So breaking things down into achievable milestones and celebrating them can go a long way in keeping the morale of the team up. Plus, when you successfully achieve multiple milestones and praise the successes of your team at each milestone, you build positive momentum for the entire team. One of the top reasons people leave a company is because they say they felt underappreciated and that their hard work went unnoticed. If you define success as overall completion and are waiting to reward the team, then you could find yourself unexpectedly trying to replace a member of your staff.

Note   Break down your project in achievable milestones that you can celebrate along the way.

I was once part of an Azure Enterprise Data Analytics build that was scoped at over a year in duration. For this project, success was determined by getting all of the data from a large number of unique data sources into a single data model for reporting. The source systems were new to the team, and the data was very complicated to grasp. We knew it was going to be challenging.

We often looked at the project and wondered how we were ever going to get through it because it seemed like we had a never-ending list of challenges. Focusing success on the end of the project left the team feeling stressed and defeated. However, by breaking the project in smaller achievable chunks, we had something to celebrate along the way. Yes, the ultimate goal still remained the same. And the project was always incomplete until all sources were in, but we had a better more positive way to get there.

So what do we mean when we say celebrate the wins? It can be anything you want it to be that acknowledges and/or rewards members of the team that helped achieve a milestone. It can be as simple as a conversation or as grand as an all-expense-paid trip to the Swiss Alps. You know your team and budget best, so do whatever you can.

If you are unsure of things you can do, here is a list of things that are great for recognizing individual or team success:

- Handwritten note

- Lunch or private meeting arranged solely to acknowledge a team or person

- Acknowledging the person or team in a company-wide communication

- A gift (if you have the means to do so)

- Day off or letting them off early

- Monetary items such as a promotion, raise, or bonus

- More responsibility or utilizing for special projects

- Team outing

A word of caution though is make sure you are evaluating team members equally and not focusing all of the attention on select employees. Don't play favorites. Showing favoritism negatively impacts the team if others feel they are deserving but were not rewarded. While everyone may not have the skills of the superstar on the team, everyone has the same ability to work hard and add value.

Have Fun

As with any big project, it's easy to get stressed and worn out along the way, so try to look for ways to have fun. We spend most of our lives working and yet so many of us do little to find ways to improve it. Most of us have to work at least until we reach retirement age, so making it as enjoyable as possible only adds to our quality of life. We understand that different companies have different cultures, but the culture is often determined by the people that work there. In fact, research shows that culture often flows outward and down the ranks from those in leadership. If you're in leadership of your Azure migration project, even informally, that means you.

If you don't like the culture, then try to influence change. In some offices we have worked in, no one had fun because they chose that, not because it was not allowed. If fun is considered taboo where you are, then you may question how healthy that place of employment is. In any job, we all have times that we have to hammer down and get our jobs done, but even in those times, try to laugh and smile along the way. Yes, you are paid to do a job, but if you are miserable doing it every day, then you are not working to your fullest potential.

Note   Company culture is determined by the people that work there.

As leaders in your organization, try to create a culture where your employees love working. Look for ways to make your office a place that people love coming to and for ways to balance work stress with some fun.

One of my favorite jobs I ever worked at was with a small startup company that really prided themselves on being an employer of choice. Fun was not only something they encouraged, but it was part of their mission statement. The office was colorful with numerous games and activities spread out throughout it. There was a stocked kitchen with food, drinks, and snacks free to all employees. Nearly every person had Nerf guns at their desk, and at any moment, all-out war would break out with the entire office.

Hard work was appreciated with rewards, and we had frequent team outings. And the holidays were massively over the top. For Halloween and Christmas, the office essentially shut down for a week for some of the most competitive and extreme office decorating you have ever seen. Going to work was not something I dreaded, but something I enjoyed. I knew my leaders cared about me and my happiness, and it showed every day. We worked hard and accomplished some amazing things during my time there, and the culture created played a big role in that for everyone.

Now we understand that this scenario is not going to be the same for everyone. Corporations often function differently than small companies or startups. However, this does not mean there are not things you can do.

Let's take a look at some ideas of things you can do to improve your office environment and have a little fun:

- Have games in the office

- Go out together

- Encourage friendships

- Decorate

- Celebrate together

- Hobbies at work

- Volunteer as a team

Creating an environment that people love working in can really set you apart from other companies. It will help you get the best of people and inspire them to always put forth their best effort. It will also encourage people to avoid leaving and influence top talent to apply to work there.

Summary

In this chapter, we looked at many factors that can help make your team more successful when engaging in a project such as a cloud database migration. With any large project that takes time and money, it is important to make sure you have business justification that supports your project and aligns with the goals of your organization. Keeping everyone supportive, involved, equipped, and informed will go a long way in the success of your project. When challenges arise and you have needs, be sure to utilize your executive sponsor to assist. Remember, they are a resource invested in your success. And never forget to celebrate the wins and have fun along the way.

CHAPTER 8

Security, Privacy, and Compliance with the Law

Once only prominent government agencies and Fortune 1000 companies paid much attention to security. Now, any IT organization with even the smallest remit must make security and legal compliance a high priority or face the consequences. The landscape for security threats is more dangerous than ever. Whereas in the past, we faced malicious or curious hackers acting alone or in small groups, today, we have entire nation-states investing significant resources, time, and energy into hacking commercial entities.

On top of that, as the value of personal and corporate data increases, the public demands greater protections for their data artifacts stored in public clouds and SaaS applications, especially social media platforms. Following that outcry from the public, more than 16 nations and legal jurisdictions passed laws to protect consumer data and prevent breaches, with more in the works. The EU led the way when it enacted the General Data Protection Regulation (GDPR) law in the early 2010s. Other nations and jurisdictions quickly followed, from Turkey's law number 6698 to South Africa's Protection of Personal Information (POPI) Act, to Japan's Act on the Protection of Personal Information (APPI), to California's California Consumer Privacy Act (CCPA).

These laws all seek to protect the data of their citizens and to incentivize public cloud providers, social media companies, and SaaS providers to protect their citizens from hacks. These laws are not mere recommendations. The laws are further backed by harsh penalties that have major consequences. Get it right or face fines that climb into the hundreds of millions of dollars or euros.

As we delve further into these topics, remember that *security* typically relates to the technical nature and to specific implementation details of the application and database

© Kevin Kline, Denis McDowell, Dustin Dorsey, Matt Gordon 2022
K. Kline et al., *Pro Database Migration to Azure*, https://doi.org/10.1007/978-1-4842-8230-4_8

you're migrating. Security covers topics like preventing unauthorized access to your systems, preventing malware, and ensuring integrity when integrating new data sources with your existing systems. No matter what country or legal jurisdiction your database and application reside, security challenges and their solutions are quite consistent.

Privacy, on the other hand, is concerned with the legal and regulatory landscape with a mind toward preventing unauthorized access to personally identifiable information (PII) as well as the prevention of tampering with or accessing PII. Thus, privacy goes hand in hand with legal compliance and can vary quite a lot from country to country or regional legal jurisdiction.

Finally, *compliance* is composed of the work processes, technology, and auditing practices that ensure IT teams are maintaining proper vigilance and practices to adequately uphold the law. Compliance is usually confirmed on a regular basis through audits, perhaps yearly, to ensure that an organization is meeting their security and privacy responsibilities.

Topics Covered in This Chapter

This chapter covers a number of lessons that are concerning Azure SQL security, which is granular in detail, and privacy, which is broader in detail. By reading this chapter, you will learn about

- Major cybersecurity frameworks

- Azure networking

- Firewalls and endpoints for Azure SQL

- AD authentication for Azure SQL

- Audit and threat detection features for Azure SQL

- Encryption, data masking, and row-level security for Azure SQL

The Threat Environment

The global pandemic of 2020 caused a remarkable acceleration in cloud adoption. In a report from Deloitte published in December 2020, organizations have increased prioritization of migration projects by more than 30% compared to adoption rates from before the pandemic. In other cases, we have seen innumerable small organizations, such as school systems and small businesses, make a hasty shift from on-premises operations to cloud first since many countries and cities mandated work-from-home (WFH) policies and broad civic shutdowns.

Unfortunately, this rapid shift to WFH especially in organizations with limited IT staffing means that hackers have more targets than ever. As a result, we have also seen a 300–400% rise in cyberattacks since the start of the global pandemic, as reported by the FBI. Similarly, McAfee reported external attacks on cloud accounts spiked well above a 600% increase during the same time period.

Organizations, large and small, have long endured a variety of cyber attacks such as DDOS and Man in the Middle, a variety of injection attacks like SQL Injection and XML Injection, and many more. However, the McAfee report shows that the global pandemic also brought a skyrocketing 800% increase in ransomware attacks, with nearly half of those attacks targeting remote desktop protocol (RDP).

Reliable vendors, like Verizon, report that around 60% of breaches succeed using techniques like hacking, social engineering (e.g., phishing and business email compromise), and security configuration errors. On top of that, research firms, like Forrester, anticipate that malicious insiders represent a growing threat in coming years, growing from 25% of all attacks to 33% over the next few years.

Use a Framework When New to Cybersecurity

If your organization is new to cybersecurity, achieving a satisfactory level of cybersecurity may seem overwhelming. The answer to best meet the requirements for security and privacy, for many organizations, is to adopt one of the popular cybersecurity frameworks. These cybersecurity frameworks provide a system of standards, guidelines, and best practices to manage risks that commonly arise in the digital world. They provide a reliable and systematic set of structures, methodologies, and workflows that organizations need to protect their important IT data and assets

In some cases, the use of cybersecurity frameworks might be mandatory, or at least strongly encouraged, for specific industries or to meet requirements in certain legal jurisdictions. For example, in order to handle credit card transactions, a business in the United States has to pass a certified audit attesting to their compliance with the Payment Card Industry Data Security Standards (PCI DSS) framework.

There are dozens of different cybersecurity frameworks, but there are three dominant frameworks that we recommend:

- The US National Institute of Standards and Technology Framework for Improving Critical Infrastructure Cybersecurity (NIST CSF), available at `www.nist.gov/cyberframework`.

- The Center for Internet Security Critical Security Controls (CIS), which includes detailed recommendations for Microsoft SQL Server, available at `www.cisecurity.org/`.

- The International Organization for Standardization (ISO) frameworks ISO/IEC 27001 and 27002, available at `www.itgovernanceusa.com/iso27002`.

The *NIST cybersecurity framework* was first intended to protect nationally important infrastructure within the United States, including assets like power plants, electricity grids, and power-generating dams. While it is broad in scope and can be quite complex, its principles are applicable wherever organizations seek improved cybersecurity. It provides ways to protect IT assets by detecting risks, responding to threats, and then recovering lost or damaged assets in the event of a security incident.

The *CIS framework for cybersecurity* is a set of continuously updated prescriptive guidance that is comprised of more than 20 "benchmarks," including benchmarks for many recent versions of Microsoft SQL Server. This is our recommended framework for organizations that don't already have a mature set of practices and/or a team specializing in IT security. Appropriately for beginners, the CIS framework offers both an "essential" set of security recommendations and an "advanced" set that provides higher levels of security, possibly at a cost of some performance.

The *ISO/IEC 27001 framework* is the international standard for cybersecurity. It requires the use of an information security management system (ISMS) capable of systematically managing an organization's information security risks, threats, and vulnerabilities. ISO 27001 requires that the organization prove to an independent auditor that it has effectively implemented and continuously operates a Plan–Do–Check–Act workflow for cybersecurity. ISO 27001 is probably overkill for any small organization that does not operate internationally.

To recap, you are not required to implement any of these cybersecurity frameworks. However, they can be very helpful for planning your own cybersecurity. And in particular, the CIS Benchmark for SQL Server provides actual T-SQL scripts to lock down a variety of potential security gaps, if you have not already conducted a security lockdown yourself.

Defend in Depth

Most database professionals are not usually tasked with cybersecurity duties. So where should we start? In the precloud computing era, the firewall was a logical place to start. It's a prominent aspect of any organization's operational IT security, and its use and application are rather obvious, just like the sturdy stone walls of a medieval castle. Is the firewall the best place to begin to secure our Azure workloads? Not really.

But let's carry the castle analogy further. Unless you've studied military history, you might not realize that there are a great deal of security features to a castle long before an attacker reaches the castle walls or the moat. For example, you don't see castles in the midst of a thick forest or at the bottom of a deep valley. That's because placing your castle at a high elevation with clean lines of sight all around enables defenders to clearly see an approaching attack and to counterattack from the safety of their walls with ease.

In the same way, the firewall is but one layer of security. The Azure SQL Database firewall only blocks or permits client requests to your databases, but it does not authenticate those requests. So let's start with a discussion of Azure networking and then examine authentication and provisioning of permissions. We will then explore aspects of the firewall for Azure SQL Database and Azure SQL Managed Instance. Then, we will close with a discussion of advanced security features available as added services in the Azure cloud.

Access Control in a Nutshell

You open a client connection with a white-listed IP address and make a call to Azure SQL Database. The Azure SQL Database firewall confirms that your client IP address can continue or, if not, it disconnects the client. Next, the client must authenticate. This stage of activity offers three methods to provide valid credentials to connect to the cloud database. First, you may use SQL Server Authentication logins, or Azure Active Directory (AAD) credentials, or Azure AD tokens. (We will cover AD authentication in greater detail later in this chapter).

The authentication step is not only important to use Azure SQL Database. It also determines your ability to access other Azure services and resources, as well as Network Address Translation (NAT). Let's start with NAT.

Network Address Translation (NAT)

Broadly speaking about networking, clients and their specific IP addresses running on a local network rarely connect directly to a service like Azure SQL Database. Instead, clients make their requests, which are passed through one or more routers, gateways, and other computers before they reach the Azure SQL Database service. Of these, the network gateway is of prime importance since it possesses a publicly addressable IP address accessible to client connections and, in turn, servers up connections to the database server(s). In most cases, the database server, such as Azure SQL Database service, sees only client IP addresses that come from the gateway.

Think of this gateway as the dispatcher at a delivery service, like at a pizza delivery shop. Many hungry customers call in, ask the dispatcher questions, and make delivery requests. The dispatcher then forwards those requests on to other staff, like delivery drivers, managers, and inventory trackers, without actually passing along the names of the original customers. If the staff needs additional information, those requests go back to the dispatcher because that is the publicly known place to inquire. The dispatcher then knows how to inquire with the consumer because they kept track of who asked for what. In the case of the Azure SQL Database firewall, the service sees only the public address, which is acting as the dispatcher.

Your gateway is even more important in cloud computing than when working within a normal enterprise network setting. Let's say you are adding a new firewall rule through the Azure Portal and you are able to see and determine the actual client IP address that the gateway sees. That is helpful only in this scenario – you are adding a rule for yourself or those who are behind the same gateway as you. But when you are adding rules for people behind other gateways, as is often the case when building or migrating an application to the Azure cloud, you'll need to find out their public IP address.

Note   You may need to use one of the many online websites, such as MxToolbox, that allow you to see public IP addresses. There are many such websites and tools. Be sure to choose an online tool that has a good reputation and is widely trusted. Also, be sure to ask the person for whom you are adding a firewall rule to visit that website to get their public IP address.

The choices you make at this point are very important to security. Why? Because all clients behind a given public IP address of a specific gateway will now be permitted past the firewall after the firewall rule is configured. When the specific gateway is already part of your organization's on-premises IT infrastructure, you are in good shape because all of the clients making requests are from your company alone. But what if one of those clients is a traveling salesperson who frequently uses tethered Wi-Fi at the coffee shop, at customer locations, or the hotel? You now have a great, big security hole. Now, any user of that salesperson's tethered Wi-Fi also has access to your corporate cloud resources. To counter this threat, first, make sure staff in this situation *always* use a VPN and, second, investigate whether it is appropriate to add other kinds of network security that segments specific services of the corporate infrastructure.

Also, keep in mind that network configuration and security can vary widely from company to company. Consequently, confirm with the appropriate IT teams to see if there are multiple IP addresses, or a range of addresses, where requests are sent across your company, for example, when your company has a secondary Internet provider, likely using multiple outbound IP addresses. In this case, you should configure rules for those multiple addresses or range of addresses where it makes sense. When in doubt, verify with your networking team

Azure Virtual Networking

Private networks within Azure environments are built upon Virtual Networks (VNets). A VNet is similar to private, on-premises networks and allows resources of different kinds to communicate. Virtual Networks span Availability Zones and provide the ability to access the Internet via external IP addresses, filter traffic by port or IP address, and integrate with Azure services. Additionally, disparate VNets may be peered together to provide routing and connectivity between environments.

Azure VNets are created with a private IP address space, and all resources created within the VNet will be assigned IP addresses from that range. VNets may be further segmented using subnets, and resources can be assigned to specific subnets. Subnets permit administrators to allocate IP addresses more efficiently and to segment resources by type, role, environment, or other attributes. Moreover, Network Security Groups can be used to secure resources within specific subnets. For instance, SQL Server VMs may be placed on a single subnet and into a Network Security Group that only allows inbound traffic on port 1433 to ensure that only SQL Server client traffic is permitted.

A SQL Managed Instance requires a subnet dedicated only to the Managed Instance with no other Azure resources in the subnet. The Managed Instance is created with a set of endpoint rules for allowing connectivity for management or by other Azure services such as application servers. Azure SQL Database is deployed using a service endpoint with settings defined when it is created to allow connectivity to the database. A firewall is used to define the IP addresses or ranges that will be permitted to connect to the database. For this reason, it is important to make sure any applications or VMs connecting to the database have static IP addresses as connectivity will fail if the IP address changes to one which is not permitted in the firewall.

Allowing Access to Services and Resources

Consider the massive array of services and products offered by the Azure public cloud. Azure offers dozens and dozens of services across the globe. All of those services and products in all of those data centers have their own IP addresses. Significantly, these IP addresses are sometimes dynamic and, if not dynamic, still change, move, or are reconfigured from time to time. Imagine the amount of time and energy you'd have to expend to stay on top of a situation like this. Just creating firewall rules for an infrastructure for your web apps, VMs, and servers would require multiple full-time workers. And that would probably only get you up to the level of barely controlled chaos.

If you use one of the GUI tools offered by Microsoft to configure your Azure SQL Database firewall, you probably noticed the option "Allow Azure services and resources to access this server." (By the way, this setting appears in the GUI tools when connectivity is allowed using a public service endpoint, which we will discuss later). You can enable or disable this powerful option by selecting either Yes or No, each choice carrying deep implications for ease of use, on the one hand, and general security, on the other.

When you select No, you tell Azure that you will manually create and configure all firewall rules. This is the more restrictive security posture and may prevent the following Azure services from working: Import/Export Service and Data Sync. You may work around these issues by using manual workarounds as described in the Azure documentation.

When you select Yes, your server allows communications from all servers and resources inside the boundaries of Azure, even parts that are not included in your Azure subscription. This is often a more permissive security posture than most enterprises want. Of course, your firewall isn't meant to be authentication. So you might still choose to enable this setting and then rely more heavily on your authentication methodology to ensure it is secure enough.

Now, let's talk about more methods of securing your Azure infrastructure, namely, Virtual Networks and Azure endpoints.

Virtual Network (VNet) Firewall Rules and Azure Private Endpoints

IP firewalls and their rules are not the end of the firewall story. After all, the IP firewall is both a form of protection by restricting access to specific IP addresses and an acknowledgment that you have exposed a public endpoint of your data to the full scope of the Internet. So what sort of options do we have when we do not wish to expose a private endpoint to the world?

If your Azure infrastructure makes heavy use of virtual machines, you might want to implement Virtual Network (VNet) firewall rules in addition to IP firewall rules in combination with Azure Private Endpoints. This combination allows you to get rid of the public, Internet-facing endpoints via a more secure set of technologies. Let's take a closer look at what Azure Private Endpoints are, how they work, and what you need to keep in mind when using them. We will then move on to Azure Secure Endpoints.

Azure Private Link

Let's first examine Azure Private Link, an Azure Networking feature available for Azure services such as Azure SQL Database, Azure Synapse Analytics, Azure Storage, Azure Key Vault, Azure Event Hubs, Azure Cosmos DB, and other services. (The complete list of covered Azure services is available at `https://docs.microsoft.com/en-us/azure/private-link/private-link-overview#availability`.)

Private Link is the service that allows you to set up private endpoints on your VNet, attaching Azure SQL Database (or any of the aforementioned Azure services) directly to your VNet with its own private IP address. Once enabled, resources within your VNet can access the Azure service, such as a VM or Azure SQL Database, through the private endpoint connection rather than going through a public endpoint.

Note   When connecting via an Azure Private Endpoint, any Azure SQL Database firewall rules are ignored. You can use one or the other, but not both when you have enabled the private endpoint. If, for some reason, you do not enable the private endpoint and allow public accessibility to the endpoint, then connections not using the private endpoint will use the normal Azure SQL Database firewall rules you have defined.

Let's explain using an analogy. Imagine you're renting multiple rooms at a motel for a big family gathering and you want to enable family members to come and go to any of the rooms as they please. Like all motels, the doors to each room open independently through a door on the outside wall(s) of the motel to open air. That means anyone can walk up to a motel room door. Not very secure! But if you want to secure these doors, all of which are open to the public, you would need to post sentries to watch all of those doors. Doing so would be a lot of difficult work.

Instead, you could rent multiple rooms at a hotel in which all rooms are accessible only through corridors within the hotel. In fact, for your family, you could rent rooms all on one corridor of the hotel. In this scenario, you would only need to watch the one set of doors for your entire family since all the rooms are on that one corridor. With only one door to watch, you have a much easier way to ensure only your family gets into the gathering.

In our analogy, the hotel is your VNet, the rooms are Azure services, your family are the users, and the private endpoint is the door connecting the hotel entrance to the corridor of rented rooms. Thus, with a private endpoint for Azure SQL Database, only your VNet resources can access your Azure SQL Database.

Setting Up a Private Endpoint

When setting up and configuring an Azure Private Endpoint, you first need a VNet and declared subnet(s) where the Azure service will attach. You will also need these three prerequisites:

1. **A defined endpoint:** The endpoint specifications including what type of Azure service you are connecting to and its IP address

2. **A Network Interface Card (NIC):** Your connection to the VNet subnet

3. **A Mapping:** A mapping of the public Fully Qualified Domain Name (FQDN) of your Azure resource to the private IP address of the private endpoint

If you use the Azure Portal to create the public endpoint, all three prerequisites deploy for you by default, creating a private DNS zone to handle the FQDN mapping for you. (You can use alternative methods like ARM Templates or PowerShell automation to define these three prerequisites manually. However, we recommend the Azure Portal since it handles all three at once.)

Clients within or connected to your VNet then use the FQDN of your Azure resource in their configuration to access resources. For example, if one of your Azure SQL Databases has the FQDN of *contoso.database.windows.net*, then all your clients would continue to use that FQDN in their connection string. Thereafter, your private DNS zone causes all traffic to that service to route through the private IP address instead, meaning that once you set up the private endpoint, you need not change your connection strings for any of your existing applications or consumers.

Restrict Public Access

Azure SQL Database, like all public Internet services, connects to the Internet by default and has a public endpoint to reach your data. Using private endpoints gives us private access to the service by mapping a private and internal IP address on our VNet to an Azure service. That means we can finally cut off the public endpoint altogether, limiting public Internet-wide access. You can see this option on Azure Portal when creating the private endpoint under the option to enable/disable the public network access under the *Firewalls and virtual networks* tab. When enabled, then only clients attached to your VNet have access to the attached databases via the private endpoint.

Note that you can use the Azure Portal to manually configure this behavior. But you have many other options available. Which to choose may depend a bit on your networking architecture and design. For example, you could limit clients to connect from more than one VNet, from on-premises networks using VNet peering, from VNet-to-VNet connections, or from VPN gateways. These options allow you to further control connections to your Azure resources across regions, clouds, and on-premises data centers and resources.

More Secure Routing

One benefit of private endpoints is that all communications go directly between clients and your logical Azure SQL Server via the private IP address so that communications no longer flow through the Azure SQL Database gateways. This bypasses your Azure SQL Database gateways such that connections always use port 1433. The benefit here is you ensure communication to your Azure SQL Database lock access down to one address on a single port from your VNet.

Connections now never go through the Azure SQL Database gateway, instead flowing through the Azure Private Endpoint. By comparison, when you use a public endpoint and an Azure SQL Database firewall, connections are first sent to one of the Azure SQL Database gateways to then get redirected back to the node hosting the database.

Private Endpoint Tips and Tricks

Here are few other things to keep in mind when using private endpoints along with Azure SQL Database:

- Private endpoints are one way. Clients within your VNet connect to the Azure SQL Database resource, or any other Azure services, but prevents Azure services from reaching into or across your VNet.

- Private endpoints apply to Azure SQL Database. But they do not apply to Azure SQL Managed Instances (MIs) because MIs are directly deployed into your VNet.

- Private endpoint acts at the logical SQL Server level. They give access to all databases on that server via the private IP address, not just a single database. Depending on your needs and the Azure services you are using, this can be a good thing or a bad thing. Keep this in mind.

- If you remove the public endpoint, also consider disabling the *Allow Azure Services* option. That way, it also blocks anything else running in an Azure data center from getting in.

- When disabling the public endpoint, also consider routinely monitoring that no Azure SQL Database firewall rules exist and the public endpoint feature remains disabled. That way, if an intruder does enable a public endpoint, there are no extant firewall rules that instantly enable and allow traffic.

- Private endpoint must exist in the same region as the VNet it attaches to, although your Azure SQL Databases may be in other regions. In addition, a single Azure SQL Database can support multiple private endpoint connections.

Azure Service Endpoints

So far, we have discussed two ways to protect your Azure SQL Database. First, we have firewall rules, a means of limiting public exposure to your database. Second, we have Azure Private Endpoints, a means to limit and control all public access, if needed. They each have their pros and cons, which could lead to some scenarios where you want a mix of the two options.

However, there is yet a third option for securing Azure SQL Database, service endpoints. For what it is worth, service endpoints are an older technology than private endpoints. It is possible that you will only need this information to maintain existing Azure infrastructures. But they also offer yet another alternative to secure your Azure infrastructure in a way that may better meet your needs.

Azure Service Endpoints are similar to private endpoints, since they functionally restrict public access by routing access through one or more of your VNet. Unlike private endpoints, service endpoints still utilize the Azure SQL Database public endpoint, but from behind the cover of the Azure SQL Database firewall. That means traffic to your Azure SQL Database or from your applications hosted in Azure might travel across public Internet routes as the traffic makes its way to the public endpoint. Another way to put it is that secure endpoints are more secure than solely using Azure SQL Database firewall rules, but less secure than private endpoints.

To implement a secure endpoint, you first create and configure your VNet and VNet rules, which define the subnet(s) allowed into a specific Azure SQL Database logical server. In this way, you create a secure pathway to access your Azure SQL Database. (Full documentation for this security technique is available at `https://docs.microsoft.com/en-us/azure/azure-sql/database/vnet-service-endpoint-rule-overview#prerequisites-1`.) Once configured, clients on the VNet, like a given process on a virtual machine (VM), access the Azure SQL Database via the VNet subnet, traverse the service endpoint, and then connect to the public endpoint of the Azure SQL Database.

Some traffic may access your Azure SQL Database through the public endpoint after passing through any Azure SQL Database firewall rules that are applied. But the VM traffic connects to the Azure SQL Database using the VM's private IP address. That way, you don't need public IP addresses for the VM or other service you want to use. If you wanted to lock access to all traffic except your VNet, you would then take the extra step of deleting all Azure SQL Database firewall rules, preventing all access through the Azure SQL Database firewall.

Creating and Using Service Endpoints

When you create a service endpoint, you also choose the type of Azure service to attach to the VNet subnet, for example, a *Microsoft.Sql* service. This also creates additional routes on the VNet subnet that direct any resource or service attached to the subnet to take a different non-Internet path when connecting to your Azure SQL Database resources, in any Azure region. If you are curious to see this in action, use the Azure Portal to connect to a VM attached to your VNet subnet, select VM NIC, then select the *Support + Troubleshooting* tab, and finally click the Effective Routes tool.

As you may have discerned, you gain some traffic efficiency because connections are routed over the Azure backbone and not across the public Internet. This routing reduces latencies while making the traffic somewhat more secure since it does not traverse public channels. You gain this benefit at the cost of lessened security because the public endpoint is still exposed and potentially vulnerable. You will still need VNet firewall rules in this scenario, as discussed earlier in the chapter.

Unlike the Azure SQL Database firewall rules, where you define public IP addresses or ranges of IP addresses, service endpoints allow you to scope access down to one or more VNet subnets. By defining VNet firewall rules in this way, DBAs can easily see the private IP address of any resources accessing the Azure SQL Databases by looking at the *client_net_address* column in the *sys.dm_exec_connections* Dynamic Management View (DMV), as shown in the following query:

```
SELECT client_net_address
FROM sys.dm_exec_connections
WHERE session_id = @@SPID
```

The preceding query will show you the IP address of a specific SQL Server session_id (also known as a spid), for example, the spid connecting a process from a specific VM within your VNet subnet. If you want to see all active connections and their IP addresses, leave off the WHERE clause. Nonsystem spids typically have a value of 51 or greater, so if you want to see all activities by user and processes like scheduled jobs, then change the WHERE clause to WHERE session_id > 50.

Keep these additional considerations in mind when using service endpoints:

- Service endpoints cannot secure access from on-premises directly.

- Service endpoints must reside in the same region as the Azure SQL Databases they access. (This restriction does not apply to every kind of Azure service.)

- DNS entries for Azure services continue to point to the public IP addresses of the service.

Choosing Between Private Endpoints and Service Endpoints

When comparing private endpoints and service endpoints, here are a few things to consider:

- Both support enabling and disabling the option *Allow Azure service and resources to access this server*.

- Both allow you to filter all forms of traffic using Azure SQL Database rules or VNet firewall rules.

- Both operate at the Azure SQL Database logical server level, not at the level of individual databases.

- Service endpoints still connect to the public endpoint of the Azure SQL Database. Private endpoints do not. Thus, you must still enable public access to route traffic to Azure SQL Databases when using service endpoints. Private endpoints instead route all traffic to the declared private endpoint on the back end, allowing you to completely disable public endpoints.

- Service endpoints are currently more widespread due to their longer time on the market. Private endpoints, on the other hand, are showing more adoption and growth in recent days.

- Service endpoints are FREE. Private endpoints charge an hourly fee for both the ingress/egress of data processed over that endpoint and for the lifetime of the private endpoint itself. (Data ingress/egress is metered separately but priced the same per unit.)

- In the odd circumstance that you have private endpoints set on a server but have also been denied public access, any declared service endpoints on the server will stop server new traffic requests. (Active connections on such a service endpoint are not terminated, but new connections are blocked.)

As mentioned earlier in this chapter, service endpoints and private endpoints support a wide variety of Azure services like Azure SQL Database. However, the lists of supported Azure services are not identical. Consequently, you should evaluate the lists of supported Azure services as you consider implementing one type of endpoint versus the other. For more information on supported Azure services, check out

- Service endpoint–supported Azure services – `https://docs.microsoft.com/en-us/azure/virtual-network/virtual-network-service-endpoints-overview`

- Private endpoint–supported Azure services – `https://docs.microsoft.com/en-us/azure/private-link/private-endpoint-overview#private-link-resource`

Authentication for Azure SQL

Azure Active Directory (AAD) is Microsoft Azure's cloud-based identity and access management service that is used to provide access to external users to Azure resources. On-premises Active Directory may be synchronized with AAD to create consistent access management across environments. Users and groups in AAD are provided access to Azure Roles and use role-based access control (RBAC) to access virtual machines and Azure services and to perform administrative tasks within an Azure tenant. Using Azure AD Sync, organizations may choose to synchronize passwords (Password Hash Synchronization) or federate their on-premises Active Directory to provide a unified login experience and prevent users from having to manage multiple passwords.

Firewalls and endpoints are only our first set in securing an Azure SQL Database or Managed Instance. Our next layer of security comes from authentication. Authentication is, of course, the means by which a user proves they are who they claim to be. Once proven, authenticated users are provided role-based security to objects and features within the database as well as row-level security within a given table. Accounts in Azure Active Directory can represent a single person, that is, a user, or a role containing many permissions assigned to many people, that is, a group. Using group accounts eases manageability by allowing you to centrally add and remove group members in Azure AD without changing detailed permissions item by item for many users or permissions in Azure SQL.

Note   The details in this section apply to Azure SQL Database, Azure SQL Managed Instance, and Azure Synapse Analytics.

When getting started with Azure SQL Database, you must create a server admin called the Active Directory administrator (AD admin) based directly upon your Azure AD tenant. Next, we will cover the two types of administrative accounts you should create when migrating from on-premises to Azure.

Required Administrator Accounts

Earlier, we mentioned that you must create an Azure AD admin for your Azure subscription account. The Azure AD admin may be either a user or group, but both cannot be configured at the same time. The AD admin creates the users and groups within a database, assigning each granular permissions appropriate to the type of work they perform on Azure SQL. You may also create an Azure SQL Database administrator, which we encourage, by providing the appropriate identity information and security provisioning. The Azure SQL Database administrator role has similar authority as the AD admin but limited to the context of Azure SQL. This arrangement is illustrated in Figure 8-1.

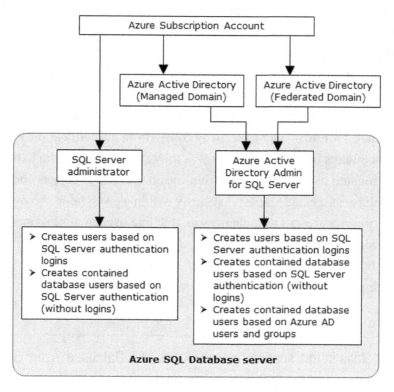

Figure 8-1. *Azure administrator roles for Azure SQL authentication management*

As mentioned earlier, AAD supports features designed to provide a single, unified login experience between on-premises and Azure services. AAD supports three core functions to accomplish this:

- Azure AD password hash synchronization is the easiest way to enable access to Azure objects for on-premises users. Users are able to use the same username and password combination in Azure as they use on-premises without additional infrastructure. Passwords are synchronized using Azure AD Connect Sync. Password hash synchronization is required for some premium Azure features.

- Azure AD Pass-through Authentication uses a lightweight software agent running on on-premises servers. Users are authenticated against on-premises Active Directory to prevent password validation from occurring in the cloud.

- When using Active Directory Federation, Azure relies on a separate, trusted system such as Active Directory Federation Services (AD FS) to validate user passwords. This method is a bit more complex to implement but supports advanced requirements such as smartcard authentication.

As the Azure AD administrator of your Azure database resources, you get the benefit of centrally managing identities and permissions of database users, in addition to all your other Azure services from the central location of the Azure Portal. Some other benefits include the following:

- Limit the proliferation of user identities across servers

- Control password rotation from a single interface

- Eliminate password storage through integrated Windows AD and Azure AD

- Support token-based authentication, as well as pass-through authentication and password hash authentication

When your Azure account is created, you are granted a default tenant for your account where you can add individual users with specific permissions to access one or more Azure services. If desired, you may create a custom default directory or even a customer domain for the default directory.

When creating new users, you should be a member of one of the admin roles just mentioned. However, other non-admin users and groups can create new users in Azure SQL if they are granted the ALTER ANY USER permission or are database users with CONTROL ON DATABASE or ALTER ON DATABASE permission for a given Azure SQL database, or are members of the db_owner database role.

Best Practices for Azure Users and Groups

Since the premise of this book is that you are migrating one or more databases from an instance of SQL Server on-premises in your data center, you probably do not need to create new users and groups or to assign traditional database permissions to them.

If in fact you do need to add new users and groups, refer to `https://azure.microsoft.com/en-us/blog/adding-users-to-your-sql-azure-database/` for step-by-step details on the process of creating users. If you need more help assigning database roles and object-level permissions, refer to `https://docs.microsoft.com/en-us/azure/azure-sql/database/logins-create-manage`.

Managed identities are the preferred way to connect applications to database services in Azure. A Managed Identity allows Azure resources to authenticate or authorize themselves with other supported Azure resources. Secrets such as database passwords are not required to be copied onto developers' machines or checked into source control. To use a Managed Identity, Azure Active Directory authentication must be configured for your Azure SQL instance. The Managed Identity is created as a contained user on the database. Ensure that a system-assigned identity is set for your app service and assign roles to the contained user. The application service is then configured to use the managed identity and connection string updated to use *Authentication=Active Directory Interactive.*

Remember that a *login* is an individual account in the master database and which links a user account in one or more user databases. A *user* is an individual account in one or more databases that may or may not be linked to a login. When a user is not linked to a login, credentials are stored with the user account. A *group* is a named collection of one or more users, with related database role memberships and object-level permissions.

We will, however, provide the most salient best practices regarding users and groups for Azure SQL:

1. Require strong passwords when using SQL Server Authentication logins. Always.

2. Better still, use AAD integration for authentication, more details on this topic appearing earlier in the chapter. This enables better leverage of managed identities in Azure. Using this approach, your applications can access your data without needing to manage and

store passwords in code, connection strings, or configurations while also reducing reliance on SQL Server Authentication accounts.

3. You must have the server-level admin account created with every Azure SQL Database as a prerequisite. Protect this account as much as possible, since it is present on every Azure SQL Database and it has the greatest amount of permissions on the server.

    a. Do not name it something obvious, like "admin" or "dba".

    b. Do not give it a weak password.

    c. Do not use it for applications to connect to a database. (We would go as far as saying "Don't give it to Dev teams.")

    d. Use it only for administrative operations. (Refer to point c.)

4. Azure AD users within a group that has the *db_owner* role assigned to it cannot issue the Create Database Scoped Credential statement in Azure SQL Database or Azure Synapse Analytics. Instead, directly grant the *db_owner* role to users who require that permission.

5. Azure SQL Managed Instances and Azure SQL DB support AAD server principals (logins). These server principals are useful to support a variety of features within Azure SQL Managed Instances, such as

    a. SQL Agent management and jobs execution

    b. Database backup and restore operations

    c. Auditing and authentication events for all statements related to AAD server principals (logins)

    d. Dedicated Administrator Connection (DAC), when granted the sysadmin server role and invoked via SSMS or the SQLCMD utility

    e. Logon triggers for logon events activated by AAD server principals (logins)

    f. Service Broker

    g. DB Mail

As mentioned earlier, you have three ways to authenticate when connecting to an Azure SQL Database: SQL Server Authentication, Azure Active Directory (AAD) credentials, and Azure AD tokens. In fact, there is also a fourth method by using Application token authentication, but that method is beyond the scope of this book. Let's now dive into the other three methods.

Connecting to Azure SQL

By this time, you should have completed the following steps to configure and utilize AAD authentication:

1. Create and populate AAD with users and groups, as needed. (Setting up AAD is beyond the scope of this book. Azure AD Connect Sync should be installed and configured to synchronize objects from an on-premises AD forest to your Azure Active Directory. Check out the vendor documentation for full details at https://docs.microsoft.com/en-us/azure/azure-sql/database/authentication-aad-configure.)

2. Accept the default directory or change the directory that is currently associated with your Azure subscription, as needed.

3. Create an Active Directory administrator (AD admin) as an individual user or group. Our advice is to create this account as a group with at least two valid users. That way, you have more than one admin in case of vacations, office emergencies, etc.

4. Configure your client computers.

5. Create contained database users in your database mapped to Azure AD identities.

6. Connect to your database by using Azure AD identity and credentials.

When users seek to authenticate to Azure SQL, they may do so through SQL Authentication, Windows Active Directory (AD), Azure Active Directory (AAD), or a hybrid of Windows AD and Azure AD authentication using Azure AD Connect.

When authenticating, you may choose among the following:

- **SQL Authentication:** Using SQL authentication to access an Azure SQL Database or Azure SQL Managed Instance is "old school." This is the easiest and most direct method of authentication because you simply provide a username and a password to authenticate using a client like SSMS. Upon creating your Azure SQL database, you must specify a server admin login with username and password. Server admins authenticate as the database owner on any database of the server or instance. The server admin will then create additional SQL logins and users, enabling them to also connect via username and password.

- **Active Directory – Password:** When users connect, they must authenticate against Azure SQL by providing a valid Azure AD identity and credentials within an AAD managed domain or federated domain.

- **Active Directory – Integrated:** This method requires that the user provide a valid domain account with the organization's on-premises Windows AD environment. This domain account must have access to Azure SQL but does not require the user to provide a password since it is validated against AAD. In this scenario, AD and AAD are synchronized using the Azure AD Connect service.

- **Active Directory – Universal with MFA Support:** This method of authentication is ideal for highly secure environments because Multi-factor Authentication (MFA) supplements the regular authentication offered by AAD. MFA requires the normal steps to sign in plus an additional layer of authentication usually by providing an access code sent directly to a given user at sign-on using SMS, a phone call, email, or another designated method.

Our recommendation is to use either AAD – Password or AAD – Integrated, whichever is most suited to your premigration plan. AAD – Universal is a stronger and more capable security model but requires some measure of extra work. Moreover, enabling Multi-factor Authentication in Azure AD provides a great deal more security with minimal effort.

Connection Tips and Tricks

Here are a few quick points to remember about Azure SQL authentication:

- The authentication methods described earlier are for use with Azure SQL Database, Azure SQL Managed Instance, and Azure Synapse Analytics. But they do not work for SQL Server running within an Azure VM. For that situation, use a Windows domain Active Directory account. Azure SQL authentication is scheduled to support Azure VMs in SQL 2022.

- The Azure AD admin for the server is the only user who may initially connect to the Azure SQL instance. They may then create and grant permissions to other subsequent Azure AD database users and groups.

- You should create the Azure SQL Server administrator preferably as an AAD group and not as an individual user administrator. The AD group should contain at least two valid users to mitigate the possibility of being locked out of the server if the DBA is on vacation or otherwise unavailable.

- You can configure either an Azure AD group or a user as an Azure SQL server admin. But you cannot create both simultaneously.

- Monitor AAD group membership changes using AAD audit activity reports. AAD authentication is also recorded in Azure SQL audit logs, but not in Azure AD sign-in logs.

- Azure RBAC permissions granted in Azure must be created and mapped manually to existing Azure SQL using existing SQL permissions, since the RBAC permissions do not apply to Azure SQL Database and Azure SQL Managed Instances.

- Old versions of SQLCMD.exe and BCP.exe do not support access to AAD. Make sure to use version 15.01 or later to use these tools with AAD.

- You must use SQLCMD version 13.1 and newer to access Azure Active Directory authentication.

- Using SSMS with AAD requires .NET Framework 4.6 or newer. If you have installed SSMS separately from the .NET Framework, you will need SSMS 2016.

- Using SSDT with AAD requires .NET Framework 4.7.2 or newer. You must use a version no older than SSDT for Visual Studio 2015 with the April 2016 update (version 14.0.60311.1). Even so, the SSDT Object Explorer cannot show Azure AD users, so query the *sys. database_principals* DMV instead when you need this information.

- Not all vendors of tools and third-party applications support Azure Active Directory.

Now that we have covered the traditional methods of security of an Azure SQL resource, from firewalls to endpoints, to Active Directory authentication, to database users and groups, let's go another level deeper with Azure SQL security features at the data layer. Remember – defend in depth!

Azure SQL Database Firewall

You might be used to securing your enterprise IT assets starting at the firewall. However, when migrating on-premises databases and applications to the Azure cloud, firewalls shrink in importance when you have a properly configured private link, as described earlier in this chapter. However, as we dive deeper, you will see that no single defensive cybersecurity feature provides full coverage. So while private links might even obviate the need for a firewall, we still recommend that you configure one in light of the principle of defense in depth. Security is now everyone's job, and we must engineer defenses in depth at every attack surface to do it well.

If you're like many IT professionals, many cybersecurity concerns sprang to the top of your mind when you first heard about cloud computing. Something along the lines of "You mean to tell me that we are going to put our data, our most precious asset, *online?!?*" Needless to say, there were (and still are) many skeptics. To use the old saying, are we just digging our own graves?

The good news is that the answer is NO. If you properly apply all of the available options for securing your data on Azure SQL Database, you can ensure your data is well protected from hackers, snoops, and even employees trying to sneak a peek at information they are not permitted to see. Later, we will address private endpoints and service endpoints. Now, let's discuss another line of defense – the Azure SQL Database firewall.

When Undesirable Clients Attempt to Connect

It's easy to tell when you displease the firewall. It will tell you something like this:

```
Cannot open server 'My_Azure_SQLServer' requested by the login. Client
with IP Address '74.120.12.123' is not allowed to access the server. To
enable access, use the Windows Azure Management Portal or run sp_set_
firewall_rule on the master database to create a firewall rules for this IP
address or address range. It may take up to five minutes for this change to
take effect.
```

As you can tell from this error message, a client whose IP address is 74.120.12.123 is attempting to access an Azure SQL Database server, but that IP address is not an acceptable value. The firewall blocks access to prohibited IP addresses or, said another way, will accept only white-listed IP addresses.

The error message goes on from there with added information. (In fact, I know more than a few cybersecurity experts who think this error message gives *too much information* to potential hackers). The added information provides details on how to fix the issue and enable access to the Azure SQL Database behind the firewall. You can simply connect to the Azure Portal and configure client IP addresses to manually fix this issue. Or you can automate this process in one of a variety of ways, like the REST API, Azure CLI, PowerShell (our favorite), or even good old Transact-SQL by invoking the sp_set_firewall_rule on the Azure SQL Database.

What Do Firewalls Do?

You can surmise from the earlier error message that the Azure SQL Database firewall restricts the range of allowable IP addresses that have permissions to connect. Naturally, you want the firewall to prevent any IP address from connecting that is outside of those commonly used by members of your organization who are appropriately cleared to access your database or databases. When a client connects to the Azure SQL Database, the IP address of the client is compared to the white-listed values in the firewall.

This security option is available for Azure SQL Database and Azure Synapse Analytics, formerly known as Azure SQL Data Warehouse. If you are running Azure SQL Database Managed Instances or SQL Server running on Azure VMs (IaaS), you will need to use alternative methods of firewalling, which we'll discuss later.

The IP address of the client is compared against not just one, but two distinct lists. The first list allows IP addresses at the server level, while the second list allows IP addresses at the database level. Note that you create only server-level firewall rules if you set the rules using Azure Portal, PowerShell, Azure CLI, or REST API–based mechanisms. However, by using Transact-SQL, you can also white-list IP addresses at the database level to connect to the Azure SQL Database.

When a client IP address attempts to connect to your Azure SQL Database, that given client's IP address value is checked first against the database-level firewall "allow list." When the client IP address finds a match on the database-level list, the client is allowed to connect. If a match is not found, the IP address value is compared to the server-level list. This behavior basically allows a client IP address two chances to find a matching IP address on one of the two white lists. If a match is not found, the client connection request is rejected.

The white list of database-level IP addresses is intended for those situations where the logically defined Azure SQL Database might be moved to another server. Because the white-listed IP addresses are stored in the database itself, it is easy to move a database elsewhere without needing to recreate the list of privileged IP addresses. The clients might need to recreate or alter their local pointer to their Azure SQL Database. But you won't have to recreate all of the firewall's allowed IP addresses. In addition, because the database-level white list is stored within the Azure SQL Database, replication mirrors the rules across replicas automatically. (Note: Azure Synapse Analytics does not offer database-level firewall rules.)

On the other hand, server-level firewall rules are very useful for administering multiple clients simultaneously who all need to connect to all of the databases on a given server, or when you have a given database (or databases) whose client IP addresses all span the same range of values. Our recommendation is to apply the most fine-grained rules you can tolerate, since it offers a significant enhancement to your organization's security.

> **Note for DBAs**   You have up to 128 server-level IP firewall rules for an Azure SQL Server. You may also have up to 128 database-level IP firewall rules, and they are allowed only on the master database and individual user databases.

> **Note for DEVs**   A common error for developers occurs when they do not specify the actual Azure SQL Database by name in the connection string. This forces the client connection to default its connection to the master database where the client does not have permissions. Make sure your connection properties include the DB name to resolve this error.

Even when you may declare server-level firewall rules, the server-level IP address values are cached at the database level to speed performance. So remember, when you remove a permitted IP address at the server level, it might reside in the cache for quite a while unless you manually run the DBCC FLUSHAUTHCACHE statement at the database level. If you want to immediately block an IP address, you must first use the KILL statement to remove the active connection from the database, then drop the IP address from the server-level white list, and then flush the authorization cache using DBCC FLUSHAUTHCACHE.

Best Practices for Firewall Rules

You've spent the last several minutes learning the concepts about Azure SQL Database firewall rules. But remember to keep the following best practices in mind when you start setting up your first Azure SQL Database:

1. Create and actively maintain standards for firewall rule names that are universally understandable and enforced throughout the organization. This best practice is made more difficult when using popular Microsoft GUI tools because they automatically create firewall rule names for you in a rather difficult format. Microsoft tools that automatically name firewall rules include Azure Data Studio, SQL Server Management Studio (SSMS), and the Azure Portal and use a convention that looks like

"ClientIPAddress_2022-4-14_14-57-10" where *ClientIPAddress* is the actual IP address of the client who created the rule. Consequently, we recommend using command-line tools, particularly PowerShell. But you can also use Azure Portal or SSMS to override the default firewall rule names. Most organizations use a standard that tells the administrator who/what the client is and where it is located, such as "London_SQLAgent_Primary".

2. Review and update your Azure SQL Database firewall rules on a regular schedule, at least once per quarter. Use PowerShell or Azure SQL Database Auditing to show all your firewall rules. Keep on the lookout not only for proper rules to be in place but also for any new rules inserted into the mix that open holes to a breach.

3. Use Azure SQL Database Auditing to capture changes to firewall rules at both the server and database level, or you can use PowerShell to return a list of the rules you have in place. Even when using alternatives to Azure SQL Database Firewall rules such as Azure Private Link, discussed later, continue to audit your firewall rules to make sure no one has opened a breach in your defenses.

4. Don't be afraid to remove firewall rules that you don't recognize or that are named contrary to your naming standards. When in doubt about a rule, remove it first and then investigate to find out how and why your team implemented that specific rule. Better yet, work with your team to maintain a change log for your Azure SQL Database firewall rules so that all rules can be fully understood and explained without a long email chain to get an answer.

5. Use database-level firewall rules as often as needed. The fine-grained security that they offer is often worth the added inconvenience. Plus, if you are maintaining a change log as recommended in the previous best practice, the script will be easy to copy, run, and modify any time you need it in the future. Remember, all connections must specify a database name.

Don't be afraid to use database-level rules, although they can only be created via T-SQL.

Viewing Azure SQL Database Firewall Rules

It's important to stay on top of your Azure SQL Database firewall rules. Typically, Azure database migration projects start out with a small number of firewall rules in place, but that number will almost assuredly grow by leaps and bounds. Here are three easy ways to see what firewall rules you have in place.

Use Azure Portal to View Firewall Rules

You can use the Azure Portal to view Azure SQL Database firewall rules. While this method is very direct, it is also a highly manual technique that may not be a good use of your time. To do so, invoke each instance of Azure SQL individually in the Portal and then select the firewall rules to view them. If you've followed our advice so far, you are presented with a list of firewall rules with easy-to-understand and consistent names. But you'll need to take special care if you have a lot of Azure SQL Databases. The more SQL databases you have, the greater the probability that you'll have exceptions to your standards. The more exceptions you have, the more difficult it becomes to maintain a consistent quality of security for all of your SQL database. Some rules might be one-offs created just to expedite a specific step in your migration, others might be transient rules that were never dropped after their usefulness had passed, and other exceptions, such as for the home offices of employees or an important remote office, might be duplicated or unaudited for long periods of time. Check for discrepancies at regular intervals, perhaps weekly.

Use PowerShell to View Firewall Rules

Our recommendation for most types of administrative work in Azure SQL Database is to use PowerShell. Compared to using the Azure Portal, you can save a great deal of work since you can easily retrieve a list of firewall rules for all servers with a single command.

You need to install the many useful Azure PowerShell cmdlets available from `https://docs.microsoft.com/en-us/powershell/azure/get-started-azureps`. We also recommend an extremely valuable collection of community-written PowerShell cmdlets at `https://dbatools.io`.

However, for this simple exercise, once you have installed the Azure PowerShell cmdlets, you will first connect and authenticate using `Connect-AzAccount` cmdlet. Your PowerShell script should look something like the following code:

```
Get-AzSqlServer `
| %{ Get-AzSqlServerFirewallRule -ServerName $_.ServerName
-ResourceGroupName $_.ResourceGroupName } `
| Select-Object -Property StartIPAddress, EndIPAddress, FirewallRuleName,
servername
```

The preceding code retrieves a list of all Azure SQL databases in the current Azure subscription. Using % as a shorthand for the cmdlet ForEachObject, the script iterates through each Azure server, executing the cmdlet Get-AzSqlServerFirewallRule as it goes. The results are then displayed for us using the Select-Object cmdlet, showing as many of the properties as we wish.

Alternatively, you can use the Cloud Shell instance in your browser to run the preceding Azure code. (Read more about the Cloud Shell instance at https://docs. microsoft.com/en-us/azure/cloud-shell/overview.)

Use Transact-SQL to View Firewall Rules

You may remember from earlier that Azure SQL Database firewall rules may be created at the server level and at the database level. While PowerShell cmdlets can only retrieve server-level firewall rules, you can use Transact-SQL to query the database-level firewall rules. In this scenario, you might simply query the DMV sys.database_firewall_rules once you have connected to and authenticated to your Azure SQL Database. You could also write a PowerShell wrapper to execute the Transact-SQL query against all instances of Azure SQL Database within your subscription.

Implementing Data Protection

Data protection is a set of features that safeguard important information from hacks, internal snooping, and inadvertent viewing by enabling encryption or obfuscation features. The data protection features we will cover were introduced piecemeal starting with SQL Server 2008's inclusion of Transparent Data Encryption (TDE) and Extensible Key Management (EKM) with other useful features released along with each new version of SQL Server. Depending on the age of the database(s) you are migrating, these features might be entirely new to you.

This section will cover data protection goals like

- Preventing high privilege but unauthorized users, like DBAs, from reading sensitive data

- Limiting data access to specific records within a table

- Encrypting data in transit so that your data is protected while it moves between client applications and the Azure SQL server

- Encrypting data at rest, when data is persisted in backup files, database files, and transaction log files

To accomplish these and other data protection goals, we will discuss Azure SQL features such as Transparent Data Encryption, Dynamic Data Masking, Row-Level Security, and Always Encrypted.

Fine-Grained Data Access Using Row-Level Security

Here's an example scenario – a small independent software vendor (ISV) contracted to develop a logistics management application for a local trucking company. The app exceeded all customer expectations and quickly earned a good reputation. Fast forward a few years to a time when the ISV now has many customers. But there is still a single database for all customers, who are kept from seeing each other's data using extensive front-end coding.

That early decision to use a single database for all customers means that the database is big, hard to manage, takes a long time to back up, and requires extensive front-end coding with each new revision just to ensure authorized users from one company cannot see the data from another company. Now that the ISV is going to migrate their SQL Server database to the cloud, they want to rearchitect their application and database to correct that technical debt. Enter row-level security (RLS).

Normally, we provide users permissions to database objects using the SQL Data Control Language (DCL) statements of GRANT, REVOKE, and DENY. But those statements only have a granularity of individual objects, like tables, views, and stored procedures. Row-level security controls the exact rows in a table that a user can access based upon characteristics of the user executing the query, like membership in a specific group.

RLS requires that you write an inline table-valued function (TVF) per database table, which is used to determine the row access logic or security predicates that control which records in the table are revealed to the user. Then, you create a security policy on top of the TVF to manage the row-level access.

The security predicate determines whether a user has read or write permissions to rows within a given table. These security predicates come in two forms. First, *filter predicates* filter rows when the user issues SELECT, UPDATE, and DELETE statements. Second, *block predicates* filter rows when user actions invoke AFTER INSERT, AFTER UPDATE, BEFORE UPDATE, and BEFORE DELETE triggers. The filter predicate ensures that the front-end application is never aware whether records were filtered or not, in some cases returning a null set if all rows are filtered. On the other hand, block predicates ensure that any operation that violates the predicate fails with an error.

Prevent Internal Snooping on Sensitive Data Using Dynamic Data Masking

Security experts say that a significant source of data breaches occur from internal staff rather than an external threat. When a DBA or System Administrator has superuser privileges on a server, for decades, there was nothing a company could do to limit their access. Now, we can put *Dynamic Data Masking* (DDM) to use. DDM is another layer on top of RLS in which we obfuscate the results of queries against sensitive data, such as credit card numbers, national ID numbers, and employee salaries. Data in the database remains unchanged, but result sets containing masked data are simply placeholders with no real value.

As an option, you could fully encrypt columns within a table. However, encryption requires a good deal of addition work with secure key management. DDM has no such requirement. We talk about encryption later in this chapter.

DDM requires that we define a mask to obfuscate the data, and then we apply the mask to one or more columns as necessary. The masks may be full, partial, or random using one of the following five types of data masks:

- **Default:** The data in the column is fully obfuscated, depending on the data type defined of the column. For example, X is used for four or less characters in a string data type, 01-01-1990 is used for dates, and <masked/> is used for XML data.

- **Email:** The data in the column is partially obfuscated except the first letter of the email address and the .com suffix of the email address. Thus, john.doe@everywhere.com becomes jXXX@XXXX.com.

- **Credit Card:** Partial masking is applied to the column showing X's to all but the last four digits of the card number.

- **Random:** The data in a column defined as a numeric data type is obfuscated by returning a randomized number extracted from a specified range which you may define.

- **Custom:** Data is partially obfuscated with a custom string, leaving only the first and last letters of the column unchanged.

DDM is limited to "normal" data types for numbers and characters. It cannot be used on encrypted columns, file streams, column sets, and sparse columns in a column set. Computed columns cannot be masked, although computed columns that depend on a masked column will return masked values. Columns used as in an index or full-text index key cannot be masked. However, if you want to mask a column used in an index, you can drop the index, apply DDM, and then recreate the index. Finally, DDM applies only at the presentation layer. That means that users with sufficient privileges to see execution plans can see unmasked data.

Protecting Data with Encryption

We have a variety of features available for the encryption of Azure SQL data, whether that data is in flight or at rest. In fact, we have several encryption options when data is at rest.

Encrypting Data in Transit Using Transport Layer Security

Data in transit from an Azure SQL Database, Azure SQL Managed Instance, or Azure Synapse Analytics server to a client application is always and inherently encrypted using Transport Layer Security (TLS). Azure SQL always encrypts all connections at all times using SSL/TLS whether or not you have enabled **Encrypt** or **TrustServerCertificate** in the application connection string. In general, you will never have to worry about whether your data in flight is properly encrypted. The answer is yes, with one exception.

The exception – some non-Microsoft drivers may not enable TLS by default or perhaps rely on an old version of TLS. If you are using a non-Microsoft driver or a very old version of TLS, you can still connect to your database, though at great risk to your security. Assuming those caveats are not in play, we recommend that you define Encrypt=True but TrustServerCertificate=False. This combination of settings forces your application to verify the server certificate, reducing your vulnerability to man in the middle attacks.

Encrypting Data at Rest Using Transparent Data Encryption

Data at rest in raw database files, transaction log files, or backups are a potential vulnerability for on-premises SQL Server as well as Azure SQL Database, Azure SQL Managed Instance, or Azure Synapse Analytics. You can mitigate these vulnerabilities by adding Transparent Data Encryption (TDE), yet another layer in your defense-in-depth planning.

TDE uses an AES encryption algorithm, AES 256, to encrypt your entire database while alleviating the need for developers to modify their existing application. The data is fully encrypted while at rest, then when needed, TDE performs real-time decryption when moving data into memory and decryption when writing data out of memory. This real-time encryption and decryption is part of database engine I/O processing at the page level.

TDE uses a symmetric key called the Database Encryption Key (DEK) to encrypt the entire database. When the database is started, TDE decrypts the DEK, making it available to decrypt and re-encrypt the database files as they are used by the various database engine processes. Also important to this process is the TDE protector, which is either the service managed certificate used to manage encryption transparently on your behalf or an asymmetric key stored in the Azure Key Vault when you need to manage the encryption keys manually. The TDE protector acts at the server level and is applied to all databases on that server for Azure SQL Database and Azure Synapse. The TDE protector is set at the instance level and inherited by all encrypted databases on that instance for Azure SQL Manage Instance.

TDE is automatically applied to all newly created databases on Azure SQL by default. TDE must be manually applied to older Azure SQL databases (pre-May 2017) and to Azure Synapse Analytics (pre-Feb 2019). These older databases and their backups are not encrypted by default. TDE then uses a built-in server certificate for the Database Encryption Key. The server certificate is maintained and rotated by the service with no input required by the users. Without the encryption key, access to the database is prevented because it cannot be decrypted and read into memory.

If you'd prefer to take direct control over the encryption keys, you can use Azure Key Vault to manage them. Azure Key Vault is a service that provides centralized key management in support of users who need Bring Your Own Key (BYOK) scenarios. Read more about BYOK at `https://docs.microsoft.com/en-us/azure/azure-sql/database/transparent-data-encryption-byok-overview`.

There are a few limitations to consider on Azure SQL Database and Azure SQL Managed Instance. For example, TDE cannot encrypt system databases, such as the master database, because it contains objects that are needed to perform TDE operations. On the other hand, database restore operations and geo-replication operations handle the encryption keys transparently. Furthermore, on Azure SQL Managed Instance, encrypted databases perpetuate their encryption status to any backups you make of them, and when you wish to restore a TDE-encrypted database, they require the TDE certificate to first be imported into the SQL Managed Instance.

Encrypting Data in Use with Always Encrypted

Cast your mind back to our discussion of Dynamic Data Masking earlier in this chapter. While DDM provides the very useful feature of transparently obfuscating columns within a table, you can instead use Always Encrypted to fully encrypt sensitive data while also allowing client applications to transparently access the data without ever revealing the encryption keys to the database engine.

The idea is to provide a degree of separation between those who own the data and may read it and those who manage the data but should not access it, such as a third-party managed services provider (MSP). Microsoft calls this concept *confidential computing.*

How Does Always Encrypted Work?

Always Encrypted was added to Azure SQL Database in 2015 and is now in all editions and service tiers and in on-premises SQL Server since SQL Server 2016. It achieves encryption quite differently from TDE. In Always Encrypted, encryption is transparent to the application but requires a driver installed on the client computer to automatically encrypt and decrypt sensitive data in the client application. The driver automatically rewrites queries in a way that semantics of the application are preserved while also encrypting or decrypting the data in sensitive columns before passing the data to and from the Database Engine.

Always Encrypted uses two types of keys: a column encryption key and a column master key. The former is used to encrypt data within a column of a table, while the latter protects one or more column encryption keys. The column encryption keys are stored as encrypted values, along with a pointer to the location of the column master key or keys, in each column of the database metadata where the Database Engine can easily access it. The column master keys are stored in trusted external key stores like Azure Key Vault and Windows Certificate Store on a client computer, or on a hardware security module.

An Always Encrypted–enabled client driver then seamlessly handles the encrypted and decrypted data stored within the encrypted column and applies the appropriate algorithm to any parameters in the SQL statements being passed between the client and server. Two types of encryption algorithms are supported:

- **Deterministic encryption:** This type of encryption always generates the same output when given the same input. It's especially useful for point lookups, equality evaluations such as for a government ID number, grouping, and indexing on encrypted columns. But because the output is consistent for a given input, hackers might be able to more easily guess certain patterns, or the real value of a given data set. This is especially true in situations where there is a limited number of possible results, such as a table containing the full name and postal code of the state, provinces, or regions of a country.

- **Randomized encryption:** This type of encryption generates a randomized output for a given column. This makes the encrypted column much more secure. But it also renders the column useless for searching, grouping, or equality evaluations for operations like joining tables. For example, you might use randomized encryption for data detailing the comments of forensic audits, which aren't used to join to other tables.

The encryption and decryption are performed by the driver on the client computer. Thus, Transact-SQL is not able to perform all actions required to configure Always Encrypted. Instead, use SSMS or PowerShell. For the same reason, some common server-side operations will not work as expected, including SELECT INTO, INSERT...SELECT, BULK INSERT, and UPDATE statements that move data from one column to another.

Always Encrypted columns have a lot of limitations, and they vary whether the column is encrypted under deterministic or randomized encryption rules. For example, columns with deterministic encryption applied require a binary2 sort order. On the other hand, Always Encrypted is not supported on columns with just about any special properties, like INDENTITY, computed columns, full-text indexes, etc.

For details about how to administer and configure Always Encrypted, read more at `https://docs.microsoft.com/en-us/sql/relational-databases/security/encryption/sql-server-encryption?view=sql-server-ver15`. If you are developing applications using Always Encrypted, consider reading all of the deep details at `https://docs.microsoft.com/en-us/sql/relational-databases/security/encryption/always-encrypted-client-development?view=sql-server-ver15`.

Advanced Data Security

While the services and features we described earlier in this chapter offer many layers to defend in depth, the features described in the next section on Advanced Data Security offer capabilities to assess, detect, categorize, and analyze your Azure data and threat environment. Another important consideration is that Advanced Data Security features are a separate and paid service from Azure SQL. It contains three distinct features: Azure Threat Detection, Data Discovery and Classification, and Vulnerability Assessment.

Advanced Threat Detection

This feature detects and alerts Azure administrators and DBAs about suspicious activities and anomalies in user patterns. The alerts provide details and potential solutions to the detected issue via integration with Azure Security Center. Advanced Threat Detection monitors your Azure SQL around the clock for hacks and unusual activity associated with hackers prepared to penetrate your systems. Azure Threat Detection is also helpful in detecting when internal personnel attempt to access data they do not usually access.

The specific threats it monitors include the following:

- **Brute-force SQL credentials attacks:** These alerts fire when an Azure SQL resource experiences an abnormally high number of failed login attempts with varying credentials.

- **Access from a potentially harmful application:** These alerts fire when a connection is made from a potentially harmful application, such as one of the common attack tools.

- **Access from an unfamiliar principal:** These alerts fire when a user logs into an Azure SQL resource using an unfamiliar or unusual SQL login.

- **Access from an unusual location:** These alerts fire when a user logs into an Azure SQL resource from a location that is different than the user's usual location.

- **Access from an unusual Azure data center:** These alerts fire when a user logs into an Azure SQL resource from a data center other than the usual or the regular data center used to log in.

- **SQL injection attacks:** SQL injection is the oldest trick in the book, and yet it is still wildly successful due to widespread poor programming practices. These alert fires when an SQL injection attack happens or if bad code is present that could result in a successful SQL injection attack.

In our opinion, these threat protections are well worth the added cost, particularly for migration projects where code that is vulnerable to SQL Injection may be lurking.

Data Discovery and Classification

At first glance, you might think that the Data Discovery and Classification (DDC) features are most relevant to data governance, not data security. However, this feature enables users to discover, classify, label, and protect sensitive data within Azure SQL. This is especially useful for data privacy and regulatory compliance scenarios, such as databases with sensitive data such as credit card numbers or financial data that must remain confidential. DDC also monitors your classified data and alerts you when that data is accessed in an unusual way.

There are two main components to DDC:

Discovery and Recommendations: The classification engine scans the Azure SQL database schemas and discovers columns that contain potentially sensitive data. It provides suggestions to protect the data and possible ways to secure the data.

Labeling: The labeling component tags columns containing sensitive data using classification metadata attributes found within the SQL engine. One of two attributes may be applied to your data – a *label*, which defines the level of sensitivity of the data, and *information types*, which further detail the type of data stored in the column. This feature is meant to aid with auditing sensitive data.

Vulnerability Assessment

As the name implies, the Vulnerability Assessment feature scans your Azure SQL databases for a wide variety of security issues, system misconfigurations, superfluous permissions, unsecured data, firewall and endpoint rules, and server-level permissions. Obviously, the Vulnerability Assessment tool is extremely valuable for detecting data security, data privacy, or data compliance issues found in a database you are migrating to Azure SQL. The Vulnerability Assessment tool uses a repository of best practices defined and updated by Microsoft, so new security issues may be added to the tool from time to time.

Summary

Securing your data is no laughing matter. Whether you are running your databases entirely within on-premises data centers or from the Azure public cloud, security should be taken more seriously than ever in the past.

## CHAPTER 9

Documenting Data Sources and Metadata in a Data Dictionary

As data ecosystems grow and become more complex, it can become difficult for organizations to understand the various datasets and what they represent. This presents challenges for data scientists and analysts who use the data to gain valuable insights. Making sense of legacy data sources in various sources requires a manual effort to understand the raw data before any value can be gained. Moreover, developers creating new features and functions need to understand the data upon which those features rely.

Enter the *Data Dictionary*. A data dictionary is akin to a traditional dictionary in that it contains names, definitions, and other attributes about the data elements in a database or information system. Additionally, it also provides and catalogs metadata or information about the data itself. Until recently, data dictionaries were commonly used in academic research projects to provide guidance on interpretation and representations.

However, as businesses have become increasingly data driven, the data dictionary has become a critical tool to help them understand their data and what it represents. In short, a data dictionary can assist organizations in avoiding inconsistencies across analytic sets, provide consistency in the collection and data usage methodologies, make data easier and faster to analyze, and enforce data standards.

When preparing to migrate your on-premises databases to Microsoft Azure, it is absolutely critical to have a complete picture of your data estate. Once built, the data dictionary can be updated on a regular schedule, and information such as row counts, data types, create and last update date, and time can all be incredibly useful as you migrate and modernize your on-premises data estate into Azure. After your migration

© Kevin Kline, Denis McDowell, Dustin Dorsey, Matt Gordon 2022
K. Kline et al., *Pro Database Migration to Azure*, https://doi.org/10.1007/978-1-4842-8230-4_9

is complete, the data catalog will be used by developers and analysts to understand and utilize your data to create new insights and features.

Ownership of the creation and administration of the data dictionary often falls within the realm of data governance. While you may not be a member of your organization's data governance team, this information is important because you will need to take time to ensure that there are no debilitating data type mismatches or hidden dependencies within the database you are migrating.

There are many third-party software applications that can be purchased and used to build a data dictionary. Additionally, Azure Data Catalog and the more recent Azure Purview provide native options with data catalog functionality. However, both of these products are rather new and do not provide all of the features you need to verify data types and dependencies in a database under migration. Consequently, we will show you both a manual process to build and assess your data dictionary as well as data governance features in the Azure products.

Creating Your Data Dictionary

Manually building a data dictionary can be a daunting and time-consuming process. Unless you are working in a purely greenfield environment, it will require a good deal of effort at the outset to perform discovery on your existing data environment. It will likely involve some manual effort for such tasks as adding extended properties. The good news is that once the initial lift is complete, the maintenance is much easier and can be automated.

Planning Your Data Dictionary

Planning a data dictionary requires understanding what elements are important to your organization and why they are important. It is important to only include elements required for your business to avoid cluttering the data dictionary with unnecessary data that can contribute to issues related to usability and administration. As is often the case, the discovery and requirements definitions may be the most time-consuming phase of the data dictionary process.

Interviews with business and technical stakeholders help to define the use cases and personas required to ensure the data dictionary provides value to the organization. Technical requirements such as data source identification and current-state evaluation to see how the datasets are being used in the existing environment can be a cumbersome but important step in building your data dictionary prior to migration to Azure.

The specific contents of a data dictionary may vary from business to business, and there are few industry standards governing the elements of a data dictionary. One data dictionary may contain a simple list of tables and other database objects, while another contains detailed metadata and descriptions of the data objects and their properties with detailed lineage and data structures. Table 9-1 describes common components of a data dictionary. The list is not to be considered comprehensive, and elements may be added or removed to meet business and technical requirements.

Table 9-1. *Common elements of a data dictionary*

Element Name	Description
Data object listing	Names and definitions of data objects
Data element properties	For example, data type, unique identifiers, size, indexes, required
Entity relationships	Data models and key relationship details
Schema details	
Row count	
Date created	
Last modified	
Entity purpose	
Data purpose	
Reference data	
Business rules	
Data lineage	
Extended properties	Extended properties include metadata added to objects to better describe their purpose

Generally speaking, elements of a data dictionary will serve one of three purposes. First, it lists elements or objects and other items within a data estate. This is an inventory or a list of items and objects for reference and to track changes, additions, and deletions over time. Next, the data dictionary shows a variety of attributes used to describe the data elements themselves. These are useful for users such as analysts and report authors to help them understand what data represents for use in reports and generating insights.

Finally, there are operational uses in which data is kept in a data dictionary and used to monitor key metrics such as usage, growth, update frequency, or deletions. Values within the data dictionary may also be used as the basis for automated actions such as sending reports or kicking off actions such as pruning databases or updates.

Extended Properties

Extended properties are an excellent way to add additional metadata to objects to increase the utility of the data element or to make it easier to understand. In Azure, extended properties are a property of objects in Azure SQL Database.

They may be added via sp_addextendedproperty, as illustrated in Listing 9-1.

Listing 9-1. Using sys.sp_addextendedproperty to add extended properties to an object

```
EXEC sys.sp_addextendedproperty
    @name=N'Revision History', @value=N'Revision number of document' ,
    @level0type=N'SCHEMA',@level0name=N'Production',
    @level1type=N'TABLE',@level1name=N'Document',
    @level2type=N'COLUMN',@level2name=N'DocumentSummary'
```

Alternatively, you can add and manipulate Extended Properties using SQL Server Management Studio (SSMS). Figure 9-1 shows the SSMS dialog screen from which that is done.

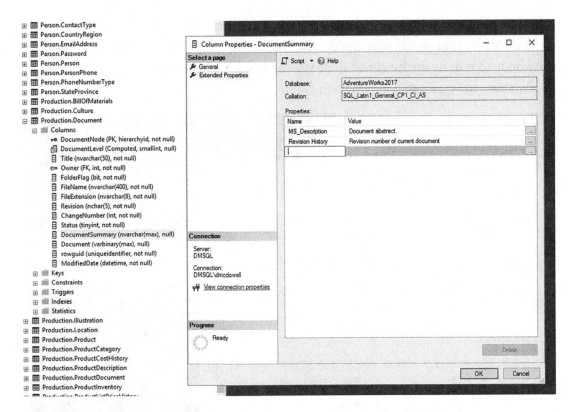

Figure 9-1. *Adding extended properties to an object using SSMS*

In Figure 9-1, we can see two extended properties for the table Production. Document on the column DocumentSummary stored in the SQL Server database AdventureWorks2017. The extended properties are MS_Description and Revision History.

Data Classification and Labels

For environments handling sensitive data such as financial or healthcare-related data, the use of data classification and data labels is a useful tool to further describe data. Data labels for sensitive data not only help an organization protect that data, but the labels are metadata that can be queried and added to the data dictionary and used to monitor access to that sensitive data and detect potential breaches.

The Data Discovery and Classification feature is available in SQL Server 2012 and later. It scans databases and provides recommendations for labeling and classifying data. Prior to SQL 2019, discovery data is stored in the properties of extended events.

In SQL Server 2019 and Azure, the data is stored in dedicated objects. Data Discovery is an SSMS feature that works with the database engine. It is lightweight in that the bulk of the workload is done on the client. In SSMS, the feature is accessed by right-clicking on the database to be assessed as shown in Figure 9-2.

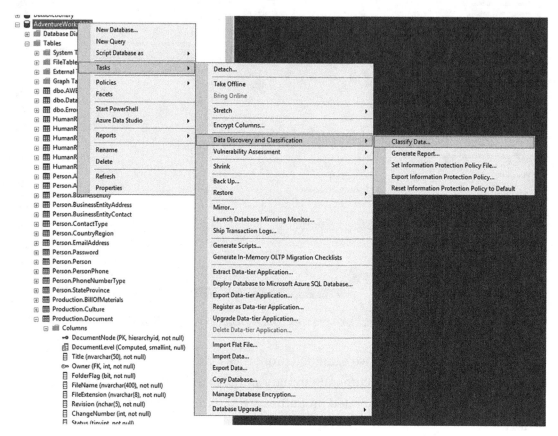

Figure 9-2. *Data Discovery and Classification in SSMS*

As you can see in Figure 9-2, you can choose to classify data within a given database, generate a report, and manage information protection policy files.

The resulting report shown in Figure 9-3 provides recommendations based upon data elements found within the database, which may require classification or protection. Recommendations may be accepted, rejected, or modified. It is also possible to manually add labels for classification.

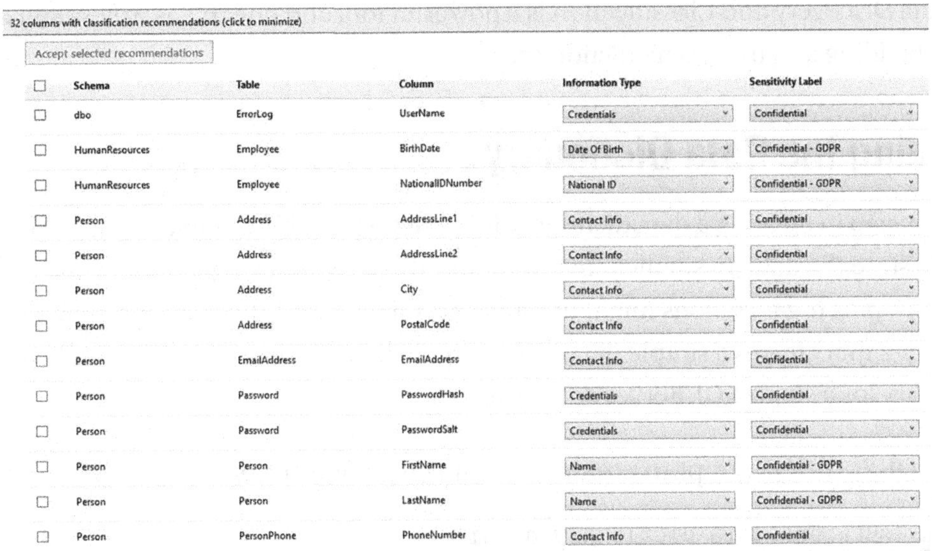

32 columns with classification recommendations (click to minimize)

Accept selected recommendations

Schema	Table	Column	Information Type	Sensitivity Label
dbo	ErrorLog	UserName	Credentials	Confidential
HumanResources	Employee	BirthDate	Date Of Birth	Confidential - GDPR
HumanResources	Employee	NationalIDNumber	National ID	Confidential - GDPR
Person	Address	AddressLine1	Contact Info	Confidential
Person	Address	AddressLine2	Contact Info	Confidential
Person	Address	City	Contact Info	Confidential
Person	Address	PostalCode	Contact Info	Confidential
Person	EmailAddress	EmailAddress	Contact Info	Confidential
Person	Password	PasswordHash	Credentials	Confidential
Person	Password	PasswordSalt	Credentials	Confidential
Person	Person	FirstName	Name	Confidential - GDPR
Person	Person	LastName	Name	Confidential - GDPR
Person	PersonPhone	PhoneNumber	Contact Info	Confidential

Figure 9-3. *SQL data classification report*

The extended attributes *sys_information_type_name* and *sys_sensitivity_label_name* are added to the selected columns and are available as attributes to be queried and stored in the data dictionary as shown in Figure 9-4.

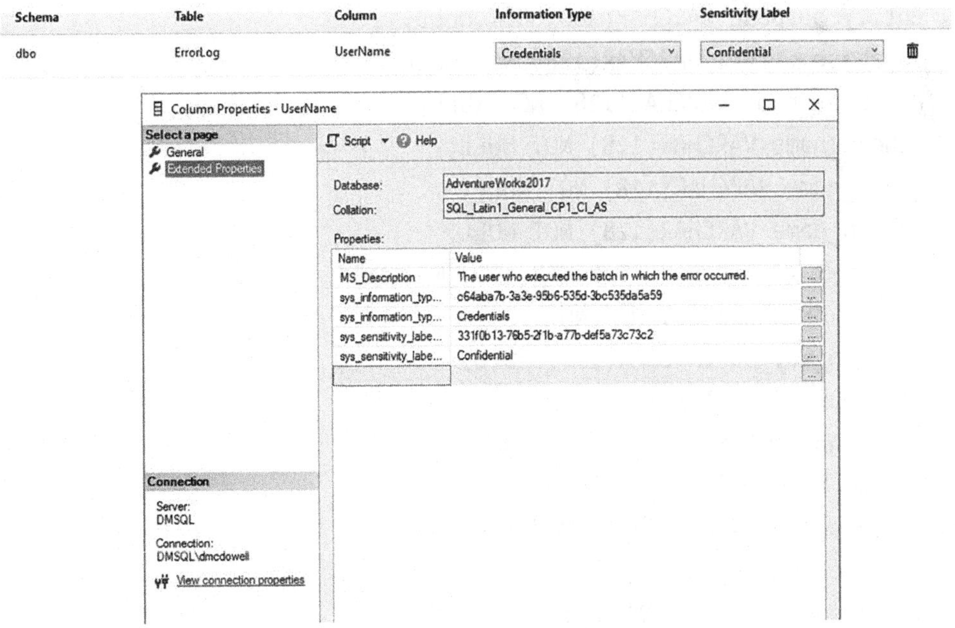

Figure 9-4. *SQL information type and information label attributes applied*

Data Discovery and Classification is a powerful tool and one that is well suited for use in building and using data dictionaries.

Creating the Data Dictionary

The first step in creating a data dictionary is to create the table or tables which will be used to store the data. In smaller environments, this may be a single table, while larger, more complex organizations might use multiple tables. Consider the script shown in Listing 9-2 that creates a simple data dictionary table which can be used to build a data dictionary for a single SQL Server database.

Listing 9-2. T-SQL script to create a simple data dictionary repository

```
CREATE TABLE dbo.data_dictionary_detail
(element_id INT IDENTITY(1,1) NOT NULL CONSTRAINT
PK_data_element_detail PRIMARY KEY CLUSTERED,
      name VARCHAR(100) NOT NULL,
      element_type VARCHAR(25) NOT NULL,
      data_type VARCHAR(128) NOT NULL,
      data_purpose VARCHAR(100) NOT NULL,
      entity_purpose VARCHAR(300) NOT NULL,
      database_server VARCHAR(128) NOT NULL,
      database_name VARCHAR(128) NOT NULL,
      schema_name VARCHAR(128) NOT NULL,
      table_name VARCHAR(128) NOT NULL,
      column_name VARCHAR(128) NOT NULL,
      original_data_source VARCHAR(500) NOT NULL,
      notes VARCHAR(1000) NOT NULL,
      foreign_key_to VARCHAR(128) NULL,
      row_count BIGINT NOT NULL,
      data_end_time SMALLDATETIME NULL,
      create_time DATETIME2(3) NULL,
      last_update DATETIME2(3) NOT NULL,
      is_deleted BIT NOT NULL,
      is_nullable BIT NOT NULL);
```

Executing the script in Listing 9-2 creates the table dbo.data_dictionary_detail, which will be populated with data dictionary elements. Each column contains data describing an element within a database within the organization's data estate. Once this table is created, you can populate it with metadata by querying SQL Server's existing system views or INFORMATION_SCHEMA views as shown in the next section.

Data Dictionary Metadata Sources

Metadata about the elements can then be inserted into the table. This can be accomplished using a number of methods. Updating the data dictionary manually would be excessively time consuming and error-prone. Instead, we recommend using a third-party product or using T-SQL scripts to build and maintain your data dictionary over time.

The Dynamic Management View (DMV) was introduced in SQL Server 2005 and has become an invaluable asset for monitoring or troubleshooting SQL Server instances. DMVs are views containing important information regarding internal SQL Server metadata and operations. DMV information includes metrics that relate to indexes, query execution, the operating system, Common Language Runtime (CLR), transactions, security, extended events, Resource Governor, Service Broker, replication, query notification, objects, input/output (I/O), full-text search, databases, database mirroring, change data capture (CDC), and other critical SQL information. These views can be queried to gain valuable insight into key performance and configuration details.

Note   Data contained within DMVs is flushed when SQL Server is restarted.

Subsequent releases of SQL Server have expanded the DMV catalog to make it a truly powerful tool for data professionals. The performance and configuration data contained within the DMVs provides an ideal source when adding usage information to your data dictionary.

In addition to DMVs, SQL Server also has its own internal version of a data dictionary in the form of system tables, also known as catalog views, like sys.objects, sys.partitions, and sys.columns. These catalog views provide a lot of information about the objects, like tables and views, within a SQL Server database and the metadata of those objects. The catalog views, however, are prone to changes and alterations as new versions of SQL Server and Azure SQL are released.

Instead, we recommend using the INFORMATION_SCHEMA schema, which contains information about the databases according to the SQL standard set of views available within a SQL instance, as well as in other Relational Database Management Systems (RDBMS) such as MySQL and PostgreSQL. The views are read only and provide access to details related to databases and objects contained on a SQL Server instance. It is automatically updated as underlying objects are changed and is used by SQL Server in query execution to validate fields, tables, and other objects referenced in a query. It is also used in query plan generation and to reference object definitions.

Note   The INFORMATION_SCHEMA, as defined in the SQL standard, is frequently referred to as "the" data dictionary across a variety of relational databases. We will use INFORMATION_SCHEMA here to avoid confusion.

Just like with DMVs, the data revealed by the INFORMATION_SCHEMA views can populate and update a data dictionary. We recommend that the INFORMATION_SCHEMA views should be used to build your data dictionary, since they are largely fixed and unlikely to change. For example, the query in Listing 9-3 retrieves from the INFORMATION_SCHEMA.COLUMNS view all user-created tables and views in the Adventureworks2017 database that contain a column called the BusinessEntityID.

Listing 9-3. Query INFORMATION_SCHEMA for all tables with a BusinessEntityID column

```
SELECT  table_name
FROM    information_schema.columns
WHERE   column_name = 'BusinessEntityID'
```

The INFORMATION_SCHEMA.COLUMNS view provides a lot more information, if we want to know more about the columns in our database tables and views. For example, there are columns in the view to show the ordinal position of the BusinessEntityID column within the table, whether it has a default value and how the default is defined, whether it is nullable, what its data type is, and so on.

Figure 9-5 shows the results in an unsorted list.

	TABLE_NAME
1	EmployeePayHistory
2	SalesPerson
3	JobCandidate
4	Password
5	SalesPersonQuotaHistory
6	Person
7	PersonCreditCard
8	vAdditionalContactInfo
9	PersonPhone
10	vEmployee
11	vEmployeeDepartment
12	vEmployeeDepartmentHistory
13	vIndividualCustomer
14	vPersonDemographics
15	vJobCandidate
16	vSalesPerson
17	SalesTerritoryHistory
18	vStoreWithDemographics
19	vStoreWithContacts
20	vStoreWithAddresses
21	vVendorWithContacts
22	vVendorWithAddresses
23	Store
24	BusinessEntity
25	BusinessEntityAddress
26	ProductVendor
27	BusinessEntityContact
28	Vendor
29	EmailAddress
30	Employee
31	EmployeeDepartmentHistory

Figure 9-5. *Query results for the INFORMATION_SCHEMA query in the AdventureWorks 2017 database from Listing 9-3*

Now, let's take the same basic form of query shown earlier in Listing 9-3 to answer a truly useful question when migrating: Which tables in our database contain columns of the same name but have different data types? Listing 9-4 shows one way to write this query.

Listing 9-4. Query INFORMATION_SCHEMA to return all schemas and tables containing columns of the same name that differ by data type

```
SELECT DISTINCT c1.column_name ,
                c1.table_schema ,
                c1.table_name ,
```

```
                    c1.data_type ,
                    c1.character_maximum_length ,
                    c1.numeric_precision ,
                    c1.numeric_scale
FROM information_schema.columns AS c1
INNER JOIN information_schema.columns AS c2
ON c1.column_name = c2.column_name
WHERE ((c1.data_type != c2.data_type)
        OR (c1.character_maximum_length != c2.character_maximum_length)
        OR (c1.numeric_precision != c2.numeric_precision)
        OR (c1.numeric_scale != c2.numeric_scale))
ORDER BY c1.column_name,
         c1.table_schema,
         c1.table_name;
```

The reason the query in Listing 9-4 is so important for a migration to Azure is that it helps reveal the potential for *implicit conversion* issues in an existing database design before you migrate the database to the cloud. Implicit conversions, as we describe later in the chapter, can cause significant performance problems and, in Azure, end up costing you a lot of additional compute dollars.

Look at the results shown in Figure 9-6.

	column_name	table_schema	table_name	data_type	character_max...	numeric_precision	numeric_scale
1	AccountNumber	Purchasing	Vendor	nvarchar	15	NULL	NULL
2	AccountNumber	Sales	Customer	varchar	10	NULL	NULL
3	AccountNumber	Sales	SalesOrderHeader	nvarchar	15	NULL	NULL
4	accountNumber	Sales	SalesOrderHeader_inMem	nvarchar	15	NULL	NULL
5	BirthDate	HumanResources	Employee	date	NULL	NULL	NULL
6	BirthDate	HumanResources	Employees	date	NULL	NULL	NULL
7	BirthDate	Sales	vPersonDemographics	datetime	NULL	NULL	NULL
8	City	HumanResources	vEmployee	nvarchar	30	NULL	NULL
9	City	Person	Address	nvarchar	30	NULL	NULL
10	City	Person	vAdditionalContactInfo	nvarchar	50	NULL	NULL

Figure 9-6. Query results from the INFORMATION_SCHEMA query returning the AdventureWorks column, schema, and table with mismatched data types

Figure 9-6 shows the first ten rows of the query result set out of a total of 133 columns. Looking at these rows, we see that the AccountNumber column appears in four tables and one of them, Sales.Customer.AccountNumber, is VARCHAR(10), which is incompatible with the NVARCHAR(15) data type of the other three tables containing a column of the same name. That means any joins between those tables will cause an implicit conversion issue. If you migrate this database to Azure as is, you would perpetuate the same database design bad practice into your Azure environment, causing you to spend more money than necessary.

Implicit conversions as described in the Microsoft documentation at `https://docs.microsoft.com/en-us/sql/t-sql/data-types/data-type-conversion-database-engine` are always something to watch for. Unlike explicit conversions in which Transact-SQL code uses the CAST or CONVERT function, implicit conversions are not visible to the user, and the output data type is determined by the SQL Server Engine.

Taken together, the information contained in DMVs and INFORMATION_SCHEMA can be used to build a robust data dictionary with valuable information for use before and after your data platform migration to Azure.

Completing the Picture

In addition to the metadata described in previous sections, there are additional items which must be well understood prior to an Azure migration. Dependencies between applications and data objects or the history of specific datasets grows more complex over time, often with little or no documentation. It is not uncommon for organizations to migrate data platforms to the cloud only to find that downstream dependencies were missed, thus causing applications to fail or perform poorly.

Moving database instances to the cloud without a clear understanding of the data lineage and relationships between instances, applications, and database objects is a common source of pain for organizations. In the following sections, we will discuss ways to discover and include these dependencies in your data dictionary for use during migration.

Linked Servers

Linked servers are a feature in SQL Server that enables you to perform distributed queries against remote database servers, that is, a "linked" server. Once you have declared a linked server, you can then execute commands against them (e.g., using OLE DB data sources) to interact with data outside of SQL Server. Typically linked servers are configured to allow the execution of Transact-SQL statements that include tables in another instance of SQL Server, a database from another vendor like Oracle, or an Azure data source like Azure Cosmos DB.

When migrating to the Azure data platform, you can transfer linked servers to Azure SQL Managed Instances, since they are supported there. But you cannot transfer linked servers to Azure SQL Database, either singleton or elastic pools. Even then, you be careful to make sure you know of any linked servers that are configured on your source database. You may have to rewrite application code to exclude or work around your linked servers if you are migrating to Azure SQL Database. Otherwise, carefully test your linked server code if on your target Azure SQL Managed Instance or SQL Server on Azure VM (IaaS) to ensure that they work as intended post migration.

To find any linked servers configured on your source SQL Server database, run the T-SQL statement EXEC sp_linkedservers to return a list of all declared external data sources. Once you have identified these servers, then run the T-SQL code shown in Listing 9-5.

Listing 9-5. Finding SQL Server code that references linked servers

```
-- For a simple result set, substitute the parameter with
-- the name of a linked server from sp_linkedservers:
SELECT OBJECT_NAME(object_id), *
FROM sys.sql_modules
WHERE definition LIKE '%MyLinkedServer%';

-- For a more detailed result set:
SELECT OBJECT_NAME (referencing_id) AS referencing_object,
referenced_server_name,
referenced_database_name,
referenced_schema_name,
referenced_entity_name
FROM sys.sql_expression_dependencies
WHERE referenced_server_name IS NOT NULL
        AND is_ambiguous = 0;
```

The queries shown in Listing 9-5 will provide ample information about stored procedures, views, user-defined functions, or any other programmable module that references a linked server. You'll need to decide from there whether to migrate the code or rewrite it depending on your target Azure SQL environment.

You can also manage linked servers in SSMS by opening the Object Explorer and then right-clicking Server Objects. From there, you can create new linked servers (right-click Server Objects, select New, and then select Linked Server) or delete an existing linked server (right-click Server Objects, right-click an existing linked server name, and then select Delete).

Object Dependencies

Dependencies between applications, databases, and instances are one of the key causes of migration failures in our experience. New applications and features are added connecting to existing data sources. Organizations grow through acquisition and either do not understand the new data architecture or implement quick integrations with their existing back-end systems. Analytic workloads require data to be moved from production instances to repositories, and documentation is sparse. In many cases, organizations take a migration approach of "we will learn the dependencies when we see what breaks" (not recommended).

Internal dependencies within a SQL Server instance must also be understood. Relationships between tables and the views and stored procedures that rely on them are critical when migrating and modernizing a data platform to Azure SQL Database or Azure Managed Instances as dependencies may break once migrated.

If you are migrating an older version of SQL Server, you can use the system stored procedure sp_depends on a database object to see its dependencies. For example, you could run it against a table to see all of the views, triggers, stored procedures, or user-defined functions that depend on it. Conversely, you could run it on a SQL module, say, a stored procedure, to see all of the tables and views that it might depend on. For example:

```
EXEC sp_depends @objname = N'AdventureWorks2012.Production.iWorkOrder' ;
```

The preceding Transact-SQL code will return all objects that depend on the trigger *iWorkOrder*.

Microsoft discourages the continued use of sp_depends going forward. For newer versions of SQL Server, use either of the system tables *sys.dm_sql_referencing_entities* or *sys.dm_sql_referenced_entities*. As you would expect, a dependency between two entities occurs when one user-defined entity, called the referenced entity, appears by name in a persisted SQL expression of another user-defined entity, called the referencing entity.

When you query either of the previously mentioned system tables, you retrieve one row per referencing or referenced entity, respectively. For syntax and examples, read more at `https://docs.microsoft.com/en-us/sql/relational-databases/system-dynamic-management-views/sys-dm-sql-referenced-entities-transact-sql`. These functions can be used to populate a data dictionary with detailed dependency information for use before and during a migration to ensure critical relationships between objects are maintained or recreated as necessary.

Whatever the reason, databases do not run in vacuums – they are used by applications to accomplish work, and those dependencies should be well understood to reduce the risk of breaking production systems after they are migrated to Azure. Understanding these dependencies means ensuring that dependent systems are either moved together or can operate in disparate environments.

Azure Service Mapping

To understand the relationship between servers and endpoints in an environment, Microsoft Azure offers Service Mapping using the Log Analytics service. This service provides the ability to map dependencies between servers, processes, inbound and outbound connection latency, and ports across any TCP-connected architecture using an agent-based mapping functionality for cloud-based and on-premises servers and instances. Service Mapping provides a near real-time picture of your environment and the network connectivity between objects, as shown in Figure 9-7.

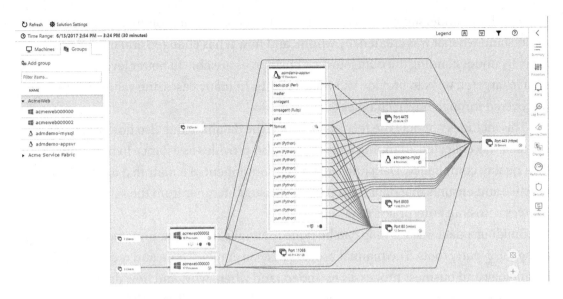

Figure 9-7. Machine groups in Azure Service Map

Service Mapping provides a dashboard with drill-down capabilities to administrators so they may understand the dependencies between endpoints in their environment.

System Mapping integrates with Azure Monitor and Operations Manager to provide additional features such as change tracking, security assessment, recommendations, and system monitoring. It also integrates with multiple third-party ITSM and monitoring platforms. During and after migration, System Mapping permits administrators to ensure that dependencies remain intact and to quickly identify issues related to connectivity between resources.

Data Lineage

Data lineage is the history of specific data across one or more database systems. Data lineage information includes where the data originates, what happens to it, and where it moves over time. For example, Point-of-Sale (POS) data might originate at POS terminals across a company's retail stores; an ETL process loads the POS flat files every night into a staging server; after applying several transformations, the POS data is then loaded into a production SQL Server database and, finally, an on-premises SQL Server Analysis Server data warehouse.

In order to understand the value of a specific set of data, it is often important to understand when it was created, by whom, and how it has changed and moved over time. By understanding the data journey, businesses are able to better leverage the data in targeted ways to provide valuable insights. In many cases, the value in data is discovered through how it changes over time.

This includes not only the records stored in the database but also how and when the data was collected. Mapping data lineage also provides assistance in managing data quality compliance. Data lineage is a key component of a mature data governance practice and ensures that data is used in a consistent manner from the same data sources across an enterprise's data ecosystem.

Building a data lineage map can be done manually or in an automated manner using third-party tools. The manual process may be cumbersome and requires both technology and business processes to understand when, why, and how data sources are added to an environment. Automating the task using tools will save time and resources while providing ongoing capture of the changes to data over time.

Azure Data Catalog

Microsoft Azure Data Catalog (ADC) is a fully managed service that provides discovery and data catalog services. To avoid confusion, we only mention Azure Data Catalog so that you can **avoid it**. ADC is entirely supplanted by Azure Purview. If you've already invested heavily in ADC, you unfortunately have no automated method to upgrade to Azure Purview. You'll have to migrate any existing data in ADC by hand or write your own scripts to move ADC data.

On the other hand, Azure Purview has excellent data discovery features. So you can avoid a lot of hand coding by simply rescanning the data sources where you want to build your data governance information.

Azure Purview

Azure Purview is a cloud-based, unified data governance solution that effectively replaces Azure Data Catalog with a more robust set of features designed to govern data across an organization's data ecosystem. Azure Purview adds features such as a unified mapping of data assets and data lineage reporting. In addition to data catalog and documentation functions, Azure Purview provides a comprehensive data governance and compliance platform that integrates natively with other Azure cloud services such as Synapse and Data Factory.

While an Azure-native or third-party solution may seem like a logical choice, there are limitations to each, including vendor lock-in, limited functionalities for certain data types and sources, and cost. For these reasons, in the following sections, we will detail ways to build your data catalog the old-fashioned way using native SQL Server methods including T-SQL scripts and DMVs.

Microsoft released Azure Purview as a consolidated data governance platform to give organizations a unified governance and management tool, enabling "single pane of glass" for governance and compliance management. Azure Purview is a powerful tool that provides much of the data discovery and cataloging functionality discussed earlier in this chapter. Azure Purview consolidates multiple tools and will be the primary hub platform for data governance and related tasks in the future.

Purview provides automated data discovery and classification as a service for data assets in an organization. This discovery creates a map of the entire data estate while integrating metadata and descriptions of the data being cataloged. When data sources are registered within Azure Purview, the data remains in the source location, while metadata is added to Azure Purview together with references to the data source location. Metadata indexes allow for easy discovery of data sources through searches.

Once registered, a data source's metadata can be enriched directly by end users. Descriptions, tags, and annotations, or even another metadata for requesting source access may be added. Enterprise users can then discover data sources that match their needs for app development, data science, business intelligence, or other uses.

The Azure Purview data map is the foundational component for discovery and governance. The map contains metadata about the analytic, Software as a Service (SaaS) data, hybrid and multicloud environments and documents their relationships across the entire data estate. Data maps are sized dynamically using storage and throughput capacity units. They scale up and down depending on storage and request load. Each capacity unit includes 25 operations per second of throughput and 10 GB of metadata storage.

Note   The Azure Purview data map is built using Apache Atlas and has a robust API library.

Data sources are registered within an Azure Purview account and scanned for object names, file sizes, columns, and other technical metadata. Additionally, schema information is extracted and classification applied for accounts connected to the Microsoft 365 Security and Compliance Center (SCC). Azure Purview provides a set of default classification rules that are automatically detected during scanning. Purview uses the same classifications, also known as sensitive information types, as Microsoft 365.

Azure Purview integrates with Microsoft Information Protection Sensitivity Label. It provides automated scanning and labeling for files in Azure Blob storage and Azure Data Lake Storage Gen 1 and Gen 2. It supports automatic labeling for database columns for SQL Server, Azure SQL Database, Azure SQL Database Managed Instance, Azure Synapse, and Azure Cosmos DB. The default classification rules are not editable, although it is possible to define your own custom classification rules using Regex or custom expressions.

You can execute scans manually or scheduled, and you can scope the scan to entire data sources or limit it to specific folders or tables. Rule sets are applied to determine the types of information a scan will discover (e.g., file types, classifications, etc.). Prepackaged rule sets are available for various data sources, or custom rule sets can be created. Metadata discovered during the scanning process is sent to the ingestion process for populating the data map. The ingestion process applies additional logic and processing to create resource sets and lineage information. The data map displays resource grouping and other metadata as shown in Figure 9-8.

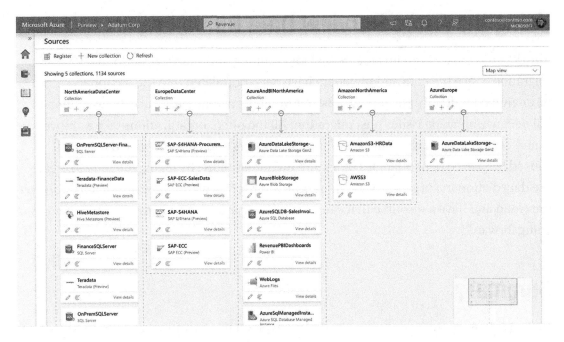

Figure 9-8. *Azure Purview data map*

The Data Catalog layer resides on top of the data map to provide search, glossary, and data lineage capabilities to the Data Insights component. The glossary is used to provide a set of terms related to the business and may contain synonyms, acronyms, and related terms. The Insight component provides access and visualizations for consumption by enterprise and clients.

Azure Purview has a powerful workflow capability that provides a set of connectors used to automate operations such as generating approval requests, validating data, and sending notifications to stakeholders within an organization. For example, the Self-Service Data Access Workflows are used to approve requests for access to data based upon rules in the policy store. Once the workflow is bound to a collection, it is automatically triggered and approves or denies access requests to data sets based upon organizational rules.

261

Using the Azure Purview Data Dictionary

The data dictionary is an invaluable tool for planning and executing a cloud data migration. Once the data and metadata have been populated into the dictionary, it can be analyzed to identify any potential issues or blockers that may cause the migration to fail or cause breaking issues post migration. All too often, organizations take a "move-it-and-see" approach with entire teams of developers and DBAs on call to fix issues caused by missed dependencies or feature incompatibility. By taking the time to create the data dictionary, organizations are setting themselves up for a successful migration to the cloud and have a resource that can be used in operations and governance practices going forward.

Summary

Having a detailed understanding of a data ecosystem not only increases the likelihood of migration success, it also helps the people and organizations who depend upon the data use it effectively. Documenting and analyzing the data and underlying metadata will give the migration team what it needs to resolve issues before they affect end users and will ensure that the data environment is well documented for the future. Once the initial heavy lift is complete, the data dictionary will continue to provide value with minimal intervention from the teams responsible for maintaining it.

CHAPTER 10

Moving Your Data to the Cloud

At this point in the book, we have looked at several considerations and factors in order to get your database to the cloud, and we are now ready to actually begin moving your data. When you start looking at how to actually migrate your data to the cloud, there are several different solutions you can use. As you should be aware by now, there is a ton of overlay with cloud technologies, and it is no different when it comes to moving your data. So some of the choice of selecting a method will come down to personal preference, past experience, or whichever tool is easiest for you. However, some options cover very specific use cases that are better suited to some situations and worse in others. Therefore, it is important to understand the available options and the pros and cons of each.

In some cases, you may be able to move your database in an all-in-one migration, meaning you move the data and schema all together, which was discussed in some detail earlier in the book. We will take another look at some of these options here and how they relate specifically to moving your data. In other cases, though, you may need or want to separate your schema and data migration. This may be true if you want to move only a subset of your data rather than the entire database or perhaps merge the data into a separate database that may already exist in the cloud.

In this chapter, we are going to review the different factors that go into making decisions on how to move your data as well as look at the different options to be able to complete those. We will not go through step by step on how to use these tools as it would likely be out of date by the time you are reading this, but we are going to look at which options are better for certain situations and things to consider to make your migration seamless and smooth. In addition, there are a lot of different options available, and we will not look at every single one but will focus on the ones most widely used.

© Kevin Kline, Denis McDowell, Dustin Dorsey, Matt Gordon 2022
K. Kline et al., *Pro Database Migration to Azure*, https://doi.org/10.1007/978-1-4842-8230-4_10

Which Service(s) Do You Need?

The first question you need to already have answered when moving your data is what service are you moving data to. In order to migrate data, you need to have a source and a target defined. In other words, you need to understand where you are at and where you are going.

Earlier in the book, you went through planning considerations that involved building your landing zone containing the service selections that compose the first draft of your Azure environment. If you skipped ahead in the book, we recommend going back and reading Chapter 2 first. As this chapter is not intended to provide a full architectural deep dive, Chapter 2 is well worth your time before proceeding further in this chapter. You should avoid going through the work of migrating your schema and data to a service only to find out afterward that a key feature you needed is not supported. This can be a waste of time and money.

Note   The first step in building a database migration plan is knowing what service you plan to migrate to and making sure it supports your needs.

This book focuses its attention on Microsoft SQL Server. However, there are several different database platforms that you can move to Azure, such as Oracle, DB2, MySQL, PostgreSQL, Cosmos DB, and various other open source and NoSQL databases. Entire books could be (and in many cases have been) written on moving any one of these databases to the cloud. These database migration scenarios, in which you migrate not only the database schema and data itself but also the entire database platform to an alternate platform, are known as *heterogeneous migration.* So there are going to be a lot of details that are not covered here, but I do want to point you to a resource that can help.

Microsoft has special documentation specifically for heterogeneous migrations called "Database Migration Guides" which you can access online at `https://docs. microsoft.com/en-us/data-migration/`. There you can find a wide variety of these migration scenarios, such as Oracle to Azure SQL Managed Instance, MySQL to Azure Database for MySQL, and many more. These guides provide best practices, guidance, and migration steps for each database type to each supported service and links to instructions on how to complete the move. If you have a specific situation not covered in this book, then this is the first place you should look.

Additionally, Microsoft offers tools specifically designed for heterogeneous migrations. When you need to migrate not only schema and data but also a lot of SQL modules like stored procedures, triggers, and user-defined functions, then the SQL Server Migration Assistant (SSMA) is the tool for you. SSMA is free and does a very good job of translating the various SQL dialects, for example, procedures written in Oracle PL/SQL to SQL Server Transact-SQL. But it is not perfect. You can usually expect 10% to 20% of the code in your heterogeneous migration to require manual conversion. Read more about SSMA at `https://docs.microsoft.com/en-us/sql/ssma/sql-server-migration-assistant`.

When you are interested in heterogeneous database migration and also require an online, continuously synchronizing migration, rather than an offline one-and-done migration, then we recommend that you use the Azure Database Migration Service (DMS). DMS is free when you use it for turnkey migrations that need 1, 2, or 4 vCores. However, DMS is not free for online migrations, which include 4 Premium vCores, but only if you use DMS longer than six months. We discuss DMS in greater detail later in this chapter.

DMS does not support as many or as diverse a set of database platforms as SSMA. But it supports the most popular platforms, including Oracle, PostgreSQL, MySQL, MongoDB (to Cosmos DB), and AWS RDS databases as well. Its main benefit, as mentioned a moment ago, is that it supports both offline and online migration scenarios for turnkey scenarios and for parallel deployment scenarios, where you need to keep both the migration source and migration target databases in use simultaneously. Read more about Azure Database Migration Services at `https://docs.microsoft.com/en-us/azure/dms/`.

Considerations on Moving Your Data

Microsoft provides us with a ton of different ways to move your data to Azure. So there are some important considerations to consider about which will work best for you. You may have shortened the list based on your service selection, but there are still many options to choose from. Now, let's take a look at some additional factors that could influence your decision.

Internet Throughput

Obviously, you'll need an Internet connection to work with your cloud assets. It is what allows us to connect, manage, and utilize Azure services and offerings. That said, the bandwidth of your connection usually only matters at the extreme ends of the network performance spectrum, for example, if you live in a region of the world with poor Internet speeds or when pushing large amounts of data to your chosen migration destination. Managing those resources can generally be done quite simply through the Azure portal or potentially via a remote desktop session on a VM.

Public cloud providers like Microsoft Azure make sure your services remain online and available at very high rates often resulting in 99.9+% uptime. For example, Azure SQL Database has a general service-level agreement (SLA) of 99.99% availability. If you make those databases zone-redundant on the Business Critical or Premium tiers, then that availability guarantee goes to 99.995%. After all, no one wants to use a cloud provider whose services constantly go offline.

But remember that the SLA is what your cloud provider offers. The speed and stability of the network service(s) you use to connect to Azure are entirely up to you. Your cloud provider does not dictate or manage your personal or corporate network. They just make sure what they offer is available. It's up to your organization or possibly you to determine the fastest and most reliable means to be able to connect.

For those who live and work near major metropolitan areas within the United States, Western Europe, Australia, South Korea, Israel, and Japan, this generally is not an issue. But for others around the world, it absolutely can be. The speed and reliability of your Internet connection to the cloud are going to be an important consideration in how you manage and utilize the cloud. And for this chapter specifically, it plays a major role in how you choose to migrate your data.

One further note is that your connection speed will continue to be a major factor if your cloud deployment is hybrid (i.e., some assets on-premises and some in the cloud). If your assets are 100% in the cloud, the need for a high-speed connection is greatly diminished. As long as you have Internet access, it is fair to say the Azure cloud will work well for you.

Note   The speed and reliability of your Internet connection are likely to be an important consideration in your database migration strategy.

While reliability is understandably important, let's look closer at the impact of our Internet speeds when undertaking a database migration. When it comes to moving data, the first choice that probably comes to mind for most is to just copy a database backup file across the Internet into the cloud-hosted environment or restore a database backup file across the network. However, depending on the size of your database and your Internet speeds, this might not be a viable option. If you have fast Internet and your database backup file is small, you are probably fine, but if your database file is large and your Internet is slow, then you have an issue. In the latter case, the file copy may take hours or even days, and, even worse, you are more likely to experience Internet outages that might result in requiring a full restart of the file copy process, leaving you back at square one.

Table 10-1 shows a chart to help you grasp the time to transmit files of varying sizes based on available network transfer speeds. It shows the estimated time required to transfer different size files across the Internet utilizing various transfer speeds that commonly exist. Please note that this is transfer speed and not Internet speeds.

Table 10-1. *Examples of time to transmit files based on transfer speeds and file size*

Transfer Speeds	10 GB File	100 GB File	1 TB File	100 TB File
10 Mbps	2 hours	22 hours	9 days	925 days
50 Mbps	25 minutes	4.5 hours	2 days	181 days
100 Mbps	13 minutes	2 hours	22 hours	92 days
500 Mbps	3 minutes	26 minutes	4 hours	4.5 days
1 Gbps	1.5 minutes	13 minutes	2 hours	9 days

Additional overhead from other processes running on your network or latency between you and the target is not factored and would further slow the file copy process. We provide the timings in Table 10-1 as estimates to help you with your decision-making. Of course, to truly understand, your network latencies require that you run your own tests.

If you want to conduct a test to measure your transfer speed to the cloud, take a small file around 1–10 GB, and then copy it into your cloud storage account while you watch the copy process with a network monitoring tool to see the actual transfer rate. You can then use that to get an idea of what your transfer rate would be like on larger volumes of data. Depending on the result, you can then determine whether you have acceptable network speeds or not.

As is expected, the larger the volume of data to migrate and the slower the transfer speed of the Internet, the longer it's going to take. For migrations that take days, a simple file copy is not realistic for most business needs. Not only based on the extremely long time span of the migration, but also because the longer the transfer time, the more likelihood of experiencing a dreaded timeout. I cannot begin to describe the scenarios throughout my career where I needed to copy data that took a large amount of time only for it to be near the end of the process, but then suddenly timeout. It is absolutely soul crushing.

Consequently, we make two recommendations in this case. First, always compress when moving large databases. Second, use the best copy tools available. This is why tools like Robocopy (short for "Robust File Copy") and AzCopy (used for copying blobs or files to or from Azure storage accounts) are critical to any part of your migration process that requires robust and resumable file copying. These tools allow multithreaded file migrations as well as restarting a file copy process from where it left off if it was interrupted.

Internet Connections

A further important consideration beyond throughput to assess is what type of Internet connection you require between your on-premises environment(s) and your cloud environment(s). The type of connection you require depends on two main factors: how securely should that data make the trip to (and potentially back from) the cloud and how acceptable is business downtime due to network latency.

Most importantly, if this data is personally identifiable information (PII), health related, finance related, or similar, it needs to be transmitted securely. That likely points you in the direction of, at a minimum, a site-to-site VPN between your data center(s) and Azure or your cloud provider of choice. Be sure to engage your security and networking team(s) for their thoughts and guidance.

When you are talking about pushing a lot of data up to the cloud and (maybe) back to on-premises, you should look in the direction of Azure ExpressRoute instead of a generic middle-tier telecom network connection. Azure ExpressRoute is a fast, private connection to Azure cloud services and is typically coordinated through your on-premises network provider. Keep in mind, however, that Azure ExpressRoute is a paid service.

Another factor when looking at the transfer time is to consider acceptable business downtime during the migration. If you have a very small window to move your data in, then a straight copy may not be a viable option if it extends anywhere near this window. For example, the business dictates that you only get two hours of downtime to move a 100 GB database. But you know from Table 10-1 that if your transfer speed is around 50 Mbps, it is going to take at least 4.5 hours. In reality, it is likely to take even longer due to line quality, error correction, and the like. At least in the scenario of meeting the business requirements, you need to consider other options. Don't worry, though, because thankfully there are other options to handle this, and we will look at them throughout this chapter.

Note   Determining acceptable business downtime is a key factor in determining what tools and methods will and will not work for your data migration.

To recap and make sure we are all on the same page, let's revisit the considerations we have looked at up until this point. First, we need to make sure we understand what service we are moving to and confirm that it supports our needs. Next, we need to look at a combination of network speed to the cloud, size of data, and available business downtime to determine if we can do a direct copy or if we need to look at other solutions.

All-in-One Tools

When you move a database, there are two things that actually get moved with it: schema and data. Your schema consists of the code and objects for your database such as the DDL for items such as tables, views, and stored procedures. These are lightweight objects that are used to bring structure to your data. Your data is the actual data being stored and what is actually consuming your storage. When factoring your data migration strategy, a consideration will need to be given to whether these two things need to migrate collectively or separately. There are tools available to just help you do one or other, and then there are tools that allow you to combine this into a single activity. In this section, we will look at some of the tools that allow you to do this all together.

Database Migration Service

The most commonly used and popular all-in-one tool for database migrations to Azure is the Database Migration Service (DMS). DMS is a Microsoft-owned tool that assists with being able to simplify your migrations, guide you through the steps of it, and automate the steps needed to complete it. It enables you to move your schema and data at scale and is optimized to be much faster than many of the other options. Commonly, this tool is used in conjunction with the Database Migration Assistant (DMA), which is an assessment and evaluation tool, but we take a look at that in greater detail in other parts of the book. Here, we are going to focus on just DMS and how to use it to move your data.

Note   The most commonly used and popular tool for database migrations is the Azure Database Migration Service (DMS).

DMS is marketed as a tool that allows you to seamlessly perform migrations from your on-premises environment to the cloud. It does this by keeping the migration simple, minimizing downtime, and by keeping things stable and secure. It not only allows you to move an entire database but also provides a lot of flexibility to separate your data and schema by selecting the objects and data you want to move rather than moving the entire database if you need to. The tool also covers a wide range of source database systems including SQL Server, MySQL, PostgreSQL, MongoDB, and Oracle to date. If you don't see your database system listed, then continue to check the official documentation because it may be something on the road map.

DMS is a completely managed service that is designed to assist you in your data migrations at scale so there are compute services involved and ultimately a cost. The pricing structure for this tool is currently broken down into two tiers: standard and premium. To understand the differences between these two, it is first important that we understand the difference between offline and online migrations. Offline migrations are often referred to as "one-time" migrations. During these migrations, the application is offline from the time you start a migration until you finish. Think you push a button to start your migration and everything is down until it is finished and you are online in the cloud. Online migrations, which are often referred to as "continuous migrations," only incur downtime when the migration completes and the actual cutover occurs. During these, you can continue to use the source database throughout the migration and only incur a downtime during the actual switch.

> **Note**   Offline migrations require downtime at the time you start, and online migrations only incur downtime at the time of the actual cutover.

As of this writing, the standard tier is free for all customers and comes in a variety of vCore options that are capped at 4 vCores and only support offline migrations. The premium tier is a paid tier that supports offline migrations with faster migration speeds, but also online migrations. If you are looking to justify the cost of the premium tier and wondering what the difference is between an online and offline migration, the difference lies in where the downtime starts. If you perform an offline migration, the database downtime (and therefore the application downtime it's supporting) begins when the migration starts. For an online migration, the downtime occurs when the database cutover occurs at the end of the migration. If you have business critical workloads you are planning to migrate and require online migrations, then it will incur a cost. For information on pricing and how you are charged, I recommend checking the latest official documentation since it will likely continue to change. If you are considering an online migration, you will also want to check the latest documentation to make sure that your database source type and target are supported since several options are not currently.

While considering those decisions, it is key to note a few prerequisites to using DMS to perform a migration. First, create an Azure Virtual Network (via VPN or ExpressRoute) that facilitates site-to-site connectivity between your on-premises source databases and the cloud-based targets. Second, ensure that port 443 is open in your network security group (NSG) rules for resources with the Service Tags of Azure Monitor, Service Bus, and Storage. Finally, if there is any firewall between your source databases and Azure, please ensure that the source databases are available through that device. As you might have guessed, this is typically a job for a networking or operations resource, although these configurations are supported by solid official tutorials and community-based blog posts if you do not have a networking person or team.

I do want to note here that while DMS provides a lot of improvement in the stability and speed in your migration, your transfer speeds are still going to limit your throughput. DMS is a great tool, but don't expect it to magically give your Internet a boost. If you have a 50 Mbps connection without DMS, you still have a 50 Mbps connection with DMS.

DMS is capable of migrating to PaaS and IaaS services, but for most database migrations involving PaaS services, this should be your first consideration. Microsoft has invested a lot of time, energy, and resources into the development of this tool to make migrations as smooth as possible. If there is a feature not listed today, I suggest continuing to check back because there are constant updates getting made here. You may also consider submitting a feature request to see if it is something that can get added to a future release.

Backup and Restore

If you have ever administered an on-premises database, typically one of the first things you learned to do is how to take a backup and perform a restore. It is not only one of the most important things we can do, but it is also one of the simplest. Yes, there are a lot of considerations given to things like backup policies, governance, and strategies that may evolve past the point of being "simple," but for most, just executing a backup and restoring it is not difficult. Whenever you perform a backup of a database, the data and schema are backed up together. Whenever you restore that backup, then likewise the schema and data are restored collectively, making this an all-in-one solution. So naturally when someone thinks of a database migration, the first thing that may come into their mind is a backup and restore. After all, it is what we do if we move a database in an on-premises environment.

In Azure, you can continue to use this method, but only for certain service types. Azure virtual machines support this in the traditional sense that many DBAs are accustomed to – being able to take a native backup directly from an on-premises database server and restore it. This makes sense because the virtual machine in the cloud is the same thing as the virtual or physical machine in your on-premises environment unless you choose to change configurations as part of your migration. The main difference is that it just lives in another data center.

For the native backup, you can take it, copy it, and restore it directly to a virtual machine pending you meet the restore requirements such as being on the same or later version of the database engine you are moving to. This is traditional and common, and many DBAs are likely familiar with this. Arguably, the preferred route for transferring backups to the cloud (especially during a migration) would be to use the BACKUP TO URL syntax. Given appropriate network permissions and connectivity, you can back up your on-premises database directly to a blog storage container. The requirements are going to be the same ones we would have to meet if we were migrating a database within our on-premises environment. The only difference here is we have to move the file to a different data center and will need to get the file to it somehow. For copying a native file to Azure, you will want to copy it to a storage account (if not backing it up to URL) or to the drives on the virtual machine to restore it.

Another option you can consider using that is exclusive to Azure virtual machines is Azure Migrate alongside Azure Site Recovery. Azure Site Recovery is a tool that lets you replicate workloads running on physical and virtual machines from on-premises to Azure. This is useful when you want to just perform a lift and shift of your existing server(s) and just continue to have them running the same way you did on-premises.

The advice of the authors of this book is that you ensure that you are not just migrating to the database destination that is comfortable for you, but one that allows you to take advantage of everything the cloud has to offer. For some specific use cases, turning toward an IaaS solution with a VM is the best decision. While the gap is rapidly closing, when you need exceptionally fast performance and log throughput, as of this writing, a well-provisioned and configured VM can still be superior to any of the PaaS offerings. That said, it is rare that lifting and shifting your on-premises environment, as is, directly to the cloud is the right answer. The PaaS database offerings include built-in high availability, managed backups and patching, and many other things that make them a compelling choice for many databases that are migrated to the cloud.

BACPAC

While you cannot restore a SQL native backup to an Azure PaaS offering, you can create a BACPAC file and import it. A *BACPAC* file is a file that can be generated that combines the data and schema from a SQL Server database in a JSON format. This method is used similarly to the backup/restore method, whereas its primary responsibility is to export and import, but it is not held to the same limitations you have with a backup. With a BACPAC file, you can import to PaaS offerings such as Azure SQL Database, whereas a native backup is limited only to Azure virtual machines. Additionally, it is important to understand that a BACPAC file is not a backup though and more of a snapshot whose primary purpose is migration. Importing a BACPAC also creates the database objects and inserts data into those objects (although it is important to keep in mind that there is significant transaction log activity associated with those inserts). BACPAC files can be created directly through SQL Server Management Studio by using the Export Data-tier Application wizard available in the database task settings. If you prefer something outside of a GUI, BACPAC files can also be created with the SqlPackage utility, PowerShell, and SQLCMD, among others.

One of the big benefits that you see with a BACPAC file is that there are no version compatibility issues, meaning you can move data and schema to different versions of SQL Server. When restoring, you will still get warnings of potential data loss due to an upgrade; however, it will provide an upgrade suggestion on how to handle it. An additional benefit is that data compression is applied, so in some cases, your BACPAC file can be smaller than your backup.

As with every suggestion, you will want to test with this approach if this is something you are considering using. There are a few potential drawbacks to this approach. First, if you are working with a large database, it can take a long time for the export and import to complete. In addition, this can put a significant load on the server while it is running, so if you are creating it on a production database, you will need to be cautious and know ahead of time what to expect. The last issue is that the tables are exported sequentially, which could cause issues with any built-in referential integrity such as foreign keys. You will not see the issue during the export, but you will encounter it during the import. My recommendation is that BACPAC files can be great, but typically best for smaller databases.

Note   BACPAC files can be great for data migrations, but typically best for smaller databases.

Log Shipping

If you need to be able to perform an online migration quickly to a virtual machine, then log shipping is a choice that you can look at. With log shipping, you restore a full backup to the SQL Server Instance running on the Azure virtual machine and then configure log shipping so that subsequent transaction logs are restored. In most cases, restoring the transaction log is going to be much faster than performing a full database restore. The cutover time is the length of time that it takes for the final transaction log backup plus the time it takes to restore it. While Microsoft doesn't use the log shipping nomenclature to describe the Log Replay Service (LRS) used to migrate to Azure SQL Managed Instance, behind the scenes, what is happening with LRS is very, very close to log shipping.

While log shipping is a SQL Server feature, other database types also have similar capabilities. If you are looking to move a different database type, then you may want to look into the options available to be able to perform this.

Physical Transfer

In situations where you have large amounts of data to transfer but are limited by your network, timing, or costs, then you can physically ship your data to Microsoft to load it on the server for you. Connecting a drive directly to the servers in an Azure Data Center and transferring the contents are going to be faster than copying data over the Internet. In fact, everyone would choose this for any moderate data size or larger if it was feasible, but the trade-off for most of us is we have to load the data from our servers, ship it to Microsoft, and have someone at Microsoft load the data, and this process can take days or even weeks. So while the unique transfer speed is faster, the process to get it there is longer. If your data is large and you estimate it is already taking a significant timeframe utilizing your existing network, you could look at shipping some or all of your data. Of course, you still have additional options such as getting additional network bandwidth, but this is oftentimes cost prohibitive and timely in its own right.

Microsoft currently has two ways you can do this. The first is the Azure Import/Export Service, which allows you to load and ship disk drives to Microsoft to load. With this option, you can choose to supply your own drives or have drives supplied by Microsoft. The other option is through Azure Data Box, which is a secure, tamper-proof, single hardware appliance that is provided by Microsoft. Microsoft sends you the appliance that you can use to load your data on and send back to them to load into Azure. Both of these options allow you to bypass the network component of the transfer that is often the biggest hurdle to getting data into Azure.

Note   For large database migrations where transferring over the network is not a viable option, you can physically transfer your data using Data Box.

There are currently three types of Azure Data Box. The first is Data Box Disk, and this is the disks you can use for the Import/Export Service should you choose to provision these from Microsoft rather than using your own. Currently, these consist of up to 5 8-TB SSD disks providing a total capacity of nearly 40 TB with each disk containing a USB\SATA interface with 128-bit encryption for data protection. Next is the standard Data Box which has a capacity of 100 TB and uses standard NAS protocols and common copy tools. This is a tough and secure device that also has AES 256-bit encryption to keep your data even safer. Finally, there is the Data Box Heavy. This beast of a device can hold up to 1 Petabyte of data and contains the same security features as the standard Data Box; the difference here is the amount of storage it has.

I have personally had to move a SQL database using a standard Data Box before. The database was so large that it would have taken us weeks to transfer across the Internet, which was not acceptable in our situation. So we ordered the data box and copied a full backup from our data center directly to the device and sent it back to Microsoft. Knowing that we also needed to perform a migration with minimal downtime, we set up differential and transaction log backups to continue and copy to the cloud as each was taken. Because these files were significantly smaller than the overall size, we could easily copy them into the cloud. Once the full database was copied and restored to the server, we were able to restore the transaction logs and differentials to get it to its current state. Once we were clear for cutover, we were able to do it in very minimal time and meet the business requirements. You will want to be sure that you have a very clear handle on how backups are happening if you do this, because if someone takes a full backup, then you break the chain. I remember we did this. I put several safeguards in place to make sure no one could take a backup on the server, and even then I constantly checked.

In some cases where it's okay to be offline for days, you can just ship it and turn it on once everything is up, but for many, you need to retain the online migration functionality, which requires a hybrid approach. It is important to note that, in the authors' experience, days of downtime for a database server is generally unacceptable. That said, in certain environments, extended downtime is permissible. In those scenarios, consider all the migration options available (and the stress on your team) and choose the one that meets commitments while keeping your team rested and focused.

User Interfaces

If you have a database backup or BACPAC file already and you just want to copy it into the cloud, then there are a couple of different ways you can do this using user interfaces. Keep in mind though that these are generally only going to be good options if your transfer speed is capable of moving the file in adequate time.

The first option here is using the Azure Portal. If you navigate to the storage account you want to copy the file to in the Azure portal, there is an option to upload your file directly into whichever directory you choose. In Figure 10-1, you can see a storage account that I have created alongside the upload option.

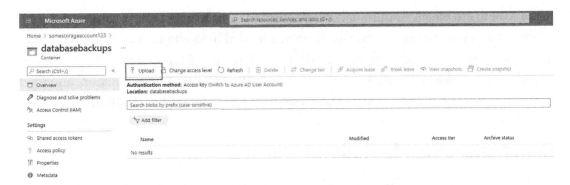

Figure 10-1. *Upload option available on an Azure storage account*

The other option to copy your file is via the Azure Storage Explorer. Azure Storage Explorer is a free tool from Microsoft that we can use to manage our storage accounts directly from our desktops. We have full control over the storage account to upload, download, and manage our data within it without needing to go to the web UI. From this tool, you can copy your database file directly into your storage account.

Handling Schema and Data Separately

In some cases, you may want to get more strategic with your data migration and not be interested in moving everything or just lifting and shifting. In these situations, you may want to consider more options that provide this by allowing you to move your schema and data separately.

Generally speaking, migrating or recreating database schema is not that hard and as time-consuming as compared to moving the data. If you have spent any time working with the database, then you probably have a good idea already of what it takes to do this. You can push from source control, use schema compare tools, or just manually script things out and move it to list a few. The part that requires a little more work is figuring out how to get the data into a new schema once you create it. In this section of the chapter, we are going to look at some of the ways you can do this.

Depending on which tools you use to move your data, you may need to land the data in a storage account first. The purpose of this is to just get the data into the cloud as simply as possible and then perform the actions you need to with it. It also mitigates your transfer across the network to one time, as opposed to needing to perform multiple times if there are issues. Additionally, if you need to transform the data in any way before loading it into your target, you will likely want it landed beforehand. This is not required for all tools, but it is something that I would give a lot of consideration to.

A lot of these tools that we will look at perform the same or similar function, and the choice on what you use often comes down to personal preference or familiarity.

Command-Line Methods

Depending on your technology background, you may be someone that loves using a graphical user interface (GUI) like a wizard in SSMS, or you may be someone that loves working with the command-line interface (CLI). Generally speaking, most people prefer using GUIs because they are easier to use, but they can oftentimes be limited by the functionality the developer created for them, whereas command-line tools can sometimes offer much greater flexibility and functionality.

The most popular command-line tool that people use with Azure is PowerShell. *PowerShell* is a Microsoft scripting language used for task automation and configuration management and is very popular among the community and Azure administrators. PowerShell has a ton of functionality that can significantly increase productivity for IT teams, and I recommend anyone getting more familiar with its capabilities. Working as an administrator in Azure, I can almost guarantee you will run across an issue at some point that can be only solved with this. In the context of this section though, PowerShell can be used to move your data. The benefit of PowerShell over other command-line tools is there is a huge repository of PowerShell script examples for Azure to help get you started.

There are, in reality, three main programmatic interfaces to the Azure services we have discussed: PowerShell, AzCLI, and the REST API. At their most basic level, they all rely on calls to the REST API, but the comfort level of teams and personnel with API calls is widely variable. That makes the other options valuable depending on the background and experience level of the people doing the work. AzCopy is a very common way to copy files to and from Azure blob storage and certainly worth an initial effort if you are copying backup files to Azure. If you have a heterogeneous OS environment, a key point to take away from this is that Azure PowerShell commands can also be run from Mac or Linux machines as well.

Azure Data Factory

In some cases, you may only need to move a subset of your on-premises data to the cloud rather than an entire database. This is a great place to consider using Azure Data Factory. Azure Data Factory (ADF) is Microsoft's fully managed cloud ETL tool that can be used to schedule and orchestrate your data move and transformation. If you are unfamiliar with the concept of ETL, it stands for Extract, Transform, and Load. This tool can be used to extract data from a source, optionally perform any needed transformations to that data, and then load it into a destination. For ADF, you have the option to load to Azure Blob storage or Azure Database.

ADF is a cost-effective tool that has a simple and code-free user interface that is easy to get up to speed on. The core component of ADF is a pipeline. A pipeline is a set of activities that defines the actions for what you need to do with a dataset. Linked services define the data source connections that ADF uses for the source and target. Together you can migrate your data using the simple built-in copy activity or apply transformation logic to get clean or model your data differently.

If you already have your own ETL tool that you are already comfortable with, then it may work as well. You will just want to make sure that it can connect to cloud resources first. As long as it does, then you should be fine using it. While Microsoft states ADF as a possible migration avenue, in the authors' experience, it is of limited utility to migrate data, and that sums up its capabilities. It is stated here because Microsoft cites it, but it is not our recommendation.

Data Synchronization/Replication

If an online migration is required, then you can consider using replication. Depending on your database type, you may have the options available to use different types of replication from your on-premises version to your cloud copy. Replication is the continuous process of copying data from a central database to other databases to keep the data synchronized. Transactional replication is the most common type of replication used for migrations, but it is asynchronous.

With all modern versions of SQL Server, AlwaysOn Availability Groups come into play as a possible migration path. While the transaction log transmission, compression, and encryption that makes AlwaysOn Availability Groups work is referred to by many as "replication," it is a far different animal than that. While replication copies the data itself (generally using data manipulation language (DML) to do that), AlwaysOn Availability

Groups compress, encrypt, and package the transaction log activity generated by database write activity and distribute that to all the replicas that are part of the availability group. While a deep dive into AlwaysOn Availability Groups is not within the scope of this book, it is important to understand two important roles that this technology may play in your migration options.

First, it is possible (with skilled network personnel) to extend the AlwaysOn Availability Group's underlying Windows Server Failover Cluster (WSFC) to Azure, place a replica in Azure, and use a failover to migrate your data to Azure. This still leaves you the task of building out a robust high availability posture in Azure but does ease the direct migration path to the cloud. The modern approach (if you are migrating from Enterprise Edition SQL Server on-premises) is to use distributed availability groups to migrate the data to Azure MI or an availability group setup within Azure itself (not recommended, as it is a dated option at this point in time). Managed Instance Link simplifies this tooling and process but is not an option if you are not migrating to Managed Instance.

Within SQL Server, you can use transactional and snapshot replication to Azure SQL Database and SQL Managed Instance to assist with your migration if you are using PaaS services. Of course, you can also use any form of vendor supported replication to an Azure VM if you are migrating to an IaaS service.

With transactional replication to a PaaS service, you set up replication similarly to how you would from an on-premises source to an on-premises target. You first set up replication using your source and target database and verify everything is working. Even though synchronous replication seems like the obvious choice, you may really want to consider if this is the right choice. In a highly transactional system configured as a synchronous availability group, you may add latency as the primary replica waits for the secondary replicas to commit the recent transactions before proceeding with the workload. If your system is low volume and some query performance degradation is acceptable, then this may be fine. You could also set up asynchronous replication, which may put the data on the replica behind, but you could take a downtime during the actual cutover, which would give the secondary copy a chance to catch up before you do the cutover. This eliminates some of the pressure on your primary database at the cost of what may be minimal extra downtime. Once you have replication setup and working, then you can redirect your application to the new copy and discontinue the replication.

Third-Party Tools and Resources

If you look at the list of options Microsoft provides and are not happy with the selection, then there are several third-party companies out in the market that offer different tools with different feature sets that you should be aware of. If you are looking for a feature that is not part of the tools or processes found in this chapter, then it may be worth searching. I am not going to list any out in this book, but you can easily discover them through your favorite search engine. Just keep in mind that if you look at alternative tools, it's often going to come at an additional cost. It will be up to you and your organization to determine whether there is value in it.

As we often suggested through this book, if you feel uncomfortable with something and are looking for additional help, reach out to Microsoft and talk with them first. They may be able to provide you the information you need to be comfortable at no cost or at a minimum refer you to a specific paid service that will help. Remember that they have a vested interest in seeing you succeed. If you abort your migration, then they are losing your business, which is something they do not want to do. However, if they are unaware of the struggles you are experiencing, then they are not going to be able to help.

If the thought of performing your own migration just feels overwhelming or if you are handicapped by staff bandwidth, then there are a lot of consulting services that specialize in database migrations that you can reach out to. These types of companies generally have staff that have gone through this process several times and should be able to assist with all aspects of the process. As expected, consulting services can be very expensive so you may want to exhaust your other options before going down this path. I always recommend starting a conversation with your cloud provider before a consulting service to see what they can offer first though. The direction still may be a consulting service, but you have looked at your options. In addition, if you do not already have a consulting service in mind, then they are great resources to provide a recommendation.

If third-party services or tools are not within your budget or timeline, do not forget that Azure Migrate (incorporating Azure Site Recovery as well) is an option. While the authors do not recommend this for regular migration of database servers, infrastructure/operations teams are likely to be fond of this option, so it behooves data teams to be familiar with it. It is not the best option for database server (and database) migration, but be open to a discussion with your networking/operations teams about all the options on the table so they are comfortable with what is executed.

Finally, there are "cloud-native" data management and backup vendors (e.g., Zerto) that offer other migration paths to Azure. While these are outside of native SQL Server and Azure pathways, keeping abreast of developments from Zerto and similar vendors will enable the data team to have productive discussions about the best migration paths for the data within the databases.

Importance of Testing

Once you have an idea of which tool or method you want to move forward with on your migration, you will want to make sure you test it as much as possible beforehand. The best approach is to go through a full migration of the data utilizing a copy of your production data to make sure that everything works as you expect it to. You will also want to test to see how much time it takes, how much downtime it will require, and ultimately if it's successful. If it does not, then you know you will need to look at other migration strategies.

Sometimes, your great ideas or what you thought may work just doesn't, and you need to pivot to a different solution. A key step in figuring this out is making sure you are testing your migrations. The last thing you want to do is be working with an untested solution during a migration only to find out something didn't work the way you expected.

Hybrid Scenarios

There may be times that you need to migrate multiple databases over to the cloud as part of a project, and it cannot be done all at once due to considerations mentioned in this chapter or due to business requirements, so you have to break it up. In these situations, you may need to run for a period of time in a hybrid environment. This means that part of the workflow may continue to live on-premises while part of it runs in the cloud. This is completely acceptable and something that often has to be done for larger and more complicated migrations. It is also why database providers will extend licensing coverage to cover both environments while your migration is in process. For example, Microsoft offers 180 days up licensing to cover both environments for the purpose of database migration.

Candidly, hybrid migration scenarios are worthy of their own book. There are a multitude of reasons why companies might want to have some of their databases on-premises and some in the cloud for a short period of time. There are important considerations for those scenarios that can impact availability, cost, and other facets of your migration. Data egress charges (executed when data leaves your Azure region of choice to be returned to web servers, app servers, or other calling services outside of that region) can be substantial and have a significant impact on the budget of your migration. There also may be technical reasons for a hybrid scenario (i.e., a legacy database needs to stay on-premises to support operations not supported in the preferred variety of Azure SQL for the project). It is important to exercise these thoroughly and lean on the Data Migration Assistant to understand what will be, and won't be, supported in your preferred cloud destination and how to accommodate that on-premises.

Summary

In this chapter, we looked at the considerations and associated methods for how to get your data to the cloud. The most important considerations when determining how to move your data are the size of the data, the transfer speed between you and the cloud, and how much downtime you have to complete the move. Once you understand those things, then you can start looking at the methods available to determine what works best. From there, you will need to determine if you want to migrate using an all-in-one tool or if you need to be able to separate your data and schema into separate migrations.

The primary tool for any database migration is going to be Data Migration Services (DMS). This is a tool that Microsoft is continually developing to make your data migrations as easy and seamless as possible. It consists of a standard tier to perform offline migrations at no cost and a premium tier that can be used for online migrations and faster offline migrations at a cost. DMS can be used to perform all-in-one migrations or used to split your data and schema up.

For all-in-one tools, you can utilize other tools such as creating a BACPAC file or utilizing log shipping for all-in-one solutions. For virtual machines, you can continue to use the backup/restore method and copy your file up or use a lift-and-shift tool such as Azure Site Recovery. Sometimes, your data is so large that transferring across the Internet is not an option, so you need to perform a physical transfer utilizing one of the Data Box options. For handling schema and data separately, you can utilize command-line tools, ETL tools, replication, or additional configuration settings using the all-in-one tools. Regardless of your need, you should be able to find a tool that helps.

CHAPTER 11

Data Validation Testing

Many organizations perform database migrations with minimal advanced planning and, similarly, do little after the migration to assure that the data, application, and server-side code work exactly as expected. It is possible to succeed when migrating a small and unsophisticated application design with this kind of "shoot from the hip" approach. But any database migration projects of significant size and complexity are always better served by going through a rigorous series of data validation tests to ensure that all aspects of your data migrated in a way that meets the organization's expectations.

While in-depth discussions on application testing are beyond the scope of this book, we will discuss methods of performing data validation testing since these project responsibilities may fall to the database professions on the migration team. In other cases, data validation testing might be a task that is set aside for subject matter experts (SMEs) from the business side of the organization, leaving the database team to focus solely on technology issues.

Having said that, the most common approach is for team members from both the business and the database sides of the house to work together on data validation testing. Therefore, the rest of this chapter will focus on common best practices in data validation under a mixed-team approach, that is, a team including members from both the business operations side of the organization and database and application development technologists from the IT side of the organization.

Topics Covered in This Chapter

Data validation is hard work. To be truly effective at validating data, the person performing the tests must possess a deep understanding of what the data within the database actually means. The testers, if they are database professionals, often know a lot about the technology issues of the data. But they are not actually able to look at a monthly business report and say "This looks right." You need business SMEs for that.

© Kevin Kline, Denis McDowell, Dustin Dorsey, Matt Gordon 2022
K. Kline et al., *Pro Database Migration to Azure*, https://doi.org/10.1007/978-1-4842-8230-4_11

The first two topics we cover are a discussion on the value of conducting data validation testing, followed by a discussion on how to best define and control the scope of the data validation task within the broader database migration effort. Finally, we delineate a number of important considerations that you should include in the various documents you produce as part of the output from the data validation task.

Keep in mind that there are many ways to conduct data validation tests. The choice that is best for you typically falls into one of three categories:

1. **Manual:** Manually performing data validation is usually time consuming and tedious. However, it has the advantage of being the most direct and expedient way to perform data validation, since you need very little preparation and can get underway immediately. Manual testing of this sort is best for small databases and applications and may not even require the participation of a database expert.

2. **Manual with sampling:** When migrating a larger database and application, the amount of data might be too big to test in meaningful ways without an enormous investment of time by the person(s) conducting the manual tests. When scaling up your data validation tests, rather than testing all data sets (e.g., in a large ERP application with over 3,000 reports), teams will often scale up their efforts by sampling the most instructive data sets from within the entirety of the database. The manual tests conducted by the team are essentially the same as a strictly manual process but are repeated for a representative sampling of the application data. For example, the database migration team might perform tests on the top 2% of customers, by revenue generated, rather than testing the reporting for every single customer of significance after a migration. From the resultant output, the migration team can effectively extrapolate the success, failure, or mixed results of their data validation tests and remediate accordingly.

3. **Automated:** While manually performing data validation tests is fine for a small project, truly large migration quality checks and tests need to be automated as much as possible if they are to be completed in a reasonable amount of time. A great deal

of data validation testing can be accomplished with a bit of clever PowerShell scripting and/or Transact-SQL programming. Automation also provides the benefit of repeatability, that is, enabling you to run the same tests many times, to run tests multiple times with a broad variety of parameters, or to run over multiple iterations when performing "what-if" analysis, all with minimal time invested.

In the following sections of this chapter, we discuss types of tests you can conduct to validate the application data, each of the approaches we've summarized earlier, and a step-by-step process you can follow to assure the business owners of the database that all migrated data is valid and ready for use.

Why Validate the Data

As we discussed in Chapter 2, proper database migration projects will include plans to validate application data as part and parcel of a successful migration. While any changes to application code or back-end Transact-SQL code should undergo their own unit tests and user-acceptance tests as a part of the migration, the data itself should also be reassessed to ensure no unexpected changes in back-end data processing have entered the picture during migration, particularly in situations where you are also upgrading versions during the course of a migration to Azure.

In addition to validating the performance and behavior of workloads on the target Azure SQL Database or Azure SQL Managed Instance, you should confirm that data is written to the database correctly and query output is as expected. By clearly defining the data validation tests for a selection of predefined input and/or outputs, you ensure that the migrated data is fit to purpose, consistent, and correct. This is important for two reasons:

1. The rate at which bugs are fixed in SQL Server has accelerated dramatically, as has the rate of cumulative updates. These changes mean that, at worst, you might experience a change in performance. However, it is not inconceivable that a fix or change could manifest in your result sets. It's better to test for full assurance that everything operates as expected. One example that precedes Azure is when Microsoft introduced a new cardinality

estimator with the release of SQL Server 2014. Many users who upgraded without testing were surprised when they experienced query regressions. Automated testing for data helps identify these changes before they become a problem.

2. Microsoft has altered the default behavior of the SQL Server engine at various times. For example, in SQL Server 2016, there were changes to the level of accuracy for values returned when there were implicit conversions between numeric and datetime types. Depending on the tolerances in your system, these variances could be something that you need to detect and then update code to maintain behavior.

Note that in these two examples, Microsoft minimizes your exposure to any risk by allowing you to use compatibility level 130. Microsoft is very diligent when it comes to documenting changes and making that information available online. It's possible to see the breaking changes to, for example, SQL Server 2016 database engine features when migrating to Azure SQL. Microsoft provides a lot of very useful information about breaking changes between version releases. Read more in this Microsoft Docs link: https://docs.microsoft.com/en-us/sql/database-engine/breaking-changes-to-database-engine-features-in-sql-server-2016?view=sql-server-ver15#SQL15.

For that reason, you should also pay close attention to the reports produced by the Azure Migration Service that describe any issues about a given database's fitness for migration. Finally, while performing these data validation tests helps to ensure all data migrated from the source to the target are correct and fit for use, that does not guarantee that the data itself is free from issues like data entry errors resulting in misspellings, transcription errors, or other clerical errors.

Scope Definition

As we have said from the outset, planning usually helps and rarely hurts on any form of project. By defining the scope and form of the data validation tests in your migration project, you have essentially created what is a "mini-planning document" for your data validation tests.

When writing your data validation scope, you may wish to include only the bare minimum – a paragraph or two summarizing the data validation tests as well as a discussion of the tests you will conduct. In addition, your scope definition should enumerate who is performing the various tests, their role (whether that be the SMEs and/or the database technologist), the deadlines for test completion, and what signifies a successful data validation test or tests.

A quick and easy series of "bare minimum" data validation tests would be to print important reports using identical parameters in the source database and the target database and compare the results. In small migration projects, it is very common for the business SME to perform this task, and when all of the numbers balance properly, they can then sign off on the data validation task of the database migration process with a high degree of confidence.

Also, in a "bare minimum" data validation test, you are most likely to test only the outputs from your migrated database that already exist, such as preexisting views and end-user reports. By comparing the outputs before and after the migrations, you will get a fairly good idea of whether the data is valid or not. (We recommend 100% accuracy between the source and the target database, with no variation between them.)

For example, you might test for

- A before and an after of each of the reports produced by the application, with consistent parameters in each test, a.k.a. "the bare minimum."

- Validate rules and constraints – in this testing scenario, you not only test the application outputs but also perform input tests to test data entry to specific data fields, ensuring that they meet your organizational standards for data types (e.g., a small integer field only accepts numeric values from -254 to +254), entity attributes (say, NULL or NOT NULL), constraint checks, foreign key checks, checks on IDENTITY values, and logical data consistency checks (e.g., the paycheck delivery date cannot precede the employee's hire date).

- Cardinality checks that count the number of records in each table and view, before and after the database migration.

- Statistical and hashing checks apply a hash calculation to important columns of a given table, such as *purchase_amount* or *order_qty*, resulting in a single calculated value per column. The resultant value might be the hashed value prior to migration (so you can compare it to the hashed value on the same column after the database migration), or one or more statistical measurements of a given column, such as AVG, SUM, and other statistical calculations.

In the following SQL code, the query retrieves a result set listing information about each table in the current database which contains columns of an integer or numeric data type:

```
SELECT table_catalog,
       table_schema,
       table_name,
       column_name,
       data_type
FROM information_schema.columns
WHERE data_type
       IN ('bigint','int','smallint','tinyint','decimal')
ORDER BY data_type;
```

The results of the preceding query are shown in Figure 11-1.

	table_catalog	table_schema	table_name	column_name	data_type
1	WideWorldImporters	Application	StateProvinces	LatestRecordedPopulation	bigint
2	WideWorldImporters	Application	StateProvinces_Archive	LatestRecordedPopulation	bigint
3	WideWorldImporters	Application	Cities	LatestRecordedPopulation	bigint
4	WideWorldImporters	Application	Cities_Archive	LatestRecordedPopulation	bigint
5	WideWorldImporters	Warehouse	ColdRoomTemperatures	ColdRoomTemperatureID	bigint
6	WideWorldImporters	Warehouse	VehicleTemperatures	VehicleTemperatureID	bigint
7	WideWorldImporters	Application	Countries	LatestRecordedPopulation	bigint

Figure 11-1. *Result set that shows database, schema, table, and column with numeric data types*

The result set shown in Figure 11-1 is not really actionable, but we can modify its query to produce an output that can be used to perform a couple quick statistical checks on each column retrieved, like this:

```
SELECT  'SELECT AVG(' + column_name + ') AS [Avg], STDEV('
     + column_name + ') AS [Stdev] FROM '
     + table_schema + '.' + table_name + ';'
FROM information_schema.columns
WHERE data_type
     IN ('bigint','int','smallint','tinyint','decimal')
ORDER BY column_name;
```

You can then copy the result set and paste it back into the SSMS Query Editor and run to derive the statistical values we mentioned earlier, as shown in Figure 11-2.

	(No column name)
1	SELECT AVG(AccountsPersonID) AS [Avg], STDEV(AccountsPersonID) AS [Stdev] FROM Sales.Invoices;
2	SELECT AVG(AlternateContactPersonID) AS [Avg], STDEV(AlternateContactPersonID) AS [Stdev] FROM Purchasing.Suppliers;
3	SELECT AVG(AlternateContactPersonID) AS [Avg], STDEV(AlternateContactPersonID) AS [Stdev] FROM Purchasing.Suppliers_Archive;
4	SELECT AVG(AlternateContactPersonID) AS [Avg], STDEV(AlternateContactPersonID) AS [Stdev] FROM Sales.Customers;
5	SELECT AVG(AlternateContactPersonID) AS [Avg], STDEV(AlternateContactPersonID) AS [Stdev] FROM Sales.Customers_Archive;
6	SELECT AVG(AmountExcludingTax) AS [Avg], STDEV(AmountExcludingTax) AS [Stdev] FROM Sales.CustomerTransactions;
7	SELECT AVG(AmountExcludingTax) AS [Avg], STDEV(AmountExcludingTax) AS [Stdev] FROM Purchasing.SupplierTransactions;

Figure 11-2. *Dynamically build a script to check for the average and standard deviation values of each numeric column in the database*

When you copy and paste the result set shown in Figure 11-2, you should execute the script in both the premigration source database and in the postmigration target database to assure the values are identical.

Data Validation Output

When performing data validation, the main goal is to create a document that proves end-user acceptance that the application data is suitable for use and fit for purpose following the database migration. Since that is the primary goal of this stage of the migration project, it is very important to create and retain at least a log of data validation tests performed and their acceptance or failure.

When you evaluate the suggested workflow for data validation, you can see why we devote quite a lot of time to discovering and detailing the database metadata elsewhere in this book. It is critical to know all of the tables and views in use in your database so that you can properly test the success of the database migration target database.

Our recommended workflow for data validation follows these steps:

1. **SME:** If the database to migrate is small, manually check all application reports for data validation before and after database migration. Be sure to use a representative set of parameter values that represent data frequently used by your end users. (Provide a checklist for the SME to sign off on each report passing the output test.)

2. **SME:** If the database to migrate is large and/or complex, manually check all application reports for data validation before and after database migration. However, your database is likely to be too big to test every report with more than one or two representative parameters. If the system is so large that manual testing might take an unreasonably long time, apply *sampling* to the reporting tests, meaning pull each report for only the one to three most important data sets represented by the report. For example, rather than running the "monthly sales report by customer" for all customers, run the report for your three most important customers. Extrapolate data quality from the sampled results. (Provide a checklist for the SME to sign off on each report passing the output test.)

3. **DBA:** Compare the cardinality of all tables and views before and after migration. Also, measure the end-to-end runtime of the queries before and after to get a rough idea of differences in processing time between source and target environment. The end-to-end runtime is, of course, not a data validation test. But if you haven't performed this test already, now is a good time to do so. (Provide a checklist for the DBA to sign off on each table passing the cardinality check.)

4. **DBA and SME:** Examine the data dictionary for tables containing "AMT" and/or "QTY" type columns, such as sales records. (Refer to Chapter 9 for more details on building your data dictionary.) Then test these columns for statistical values like AVG, SUM, and PERCENTILE_COUNT (i.e., the median value) for such columns

before and after migration. Also, measure the end-to-end runtime of the queries before and after to get a rough idea of differences in processing time between source and target environment. (Provide a checklist for the SME to sign off on each report passing the output test.)

Note that it is quite easy for a SQL developer or DBA of intermediate skill to create Transact-SQL scripts that can perform steps 3 and 4 in an automated fashion. That makes it very easy to run these quantitative data validation tests as much as needed.

Depending on how application reports are physically created, managed, and distributed, steps 1 and 2 may (or may not) be easy to automate. For example, these steps may be quite easy to automate if all of your application reports are in SQL Server Reporting Services (SSRS). But automation might be more difficult when all of your reports are in a third-party product such as Tableau, Qliq, and others or, even worse, when your application database uses multiple kinds of reporting services.

Ideally, all of your application reports are written as stored procedures, making it easy to create a Transact-SQL script to automatically pull all of the reports, with as few or as many parameters as you like. You may have noticed that in steps 3 and 4, we recommend measuring the elapsed time for each data validation test. The reason for this is to provide additional useful information about your database migration.

Why? As it turns out, many migration projects do a good job of planning their database migrations but are primarily focused on feature parity between their on-premises database applications with the same application migrated to Azure. However, few teams remember to spend any time testing for adequate performance of their application after moving to the cloud.

In turn, the migration efforts are functionally successful with the migration but experience big disappointments when balancing performance in the cloud with costs for a given level of performance. Since performance testing wasn't a major consideration during migration, they later discover that they don't have enough resources in their cloud landing zone to provide adequate performance. Consequently, the application users have to put up with suboptimal performance to stay within their budgets or choose to overspend their budgets to achieve an adequate level of performance.

That is why, in this specific case, we recommend that DBAs and other database professionals assisting with data validation testing also measure the elapsed time when running each report or when querying views and then compare the before-migration runtimes against the after-migration runtimes. When not a full database performance benchmark, these metrics provide a kind of "quick and dirty" measurement for the migration team to confirm whether performance is within expected boundaries.

Summary

It's very important to conduct data validation testing when migrating a SQL Server database to the Azure cloud. Your testing assures the organization's end users that all of the migrated data is fit for use, complete, consistent, and accurate. During this testing process, you should start by defining the scope of the tests, their full timeline, and all of the database objects that you will test, for example, end-user reports, important and frequently used queries, and views. Your tests will likely include a large number of "before-migration" and "after-migration" comparisons to make sure all of the data in the migrated application match.

If manually testing all of these important components would take too much time, you can investigate time-saving alternatives such as applying a sampling frequency when performing your tests and extrapolating those results to the entire database to be migrated, or automating parts or the entirety of your data validation tests. Finally, unless there is another step in your migration in which you will test overall application performance before and after the migration, then include measurements of the elapsed time of each data validation test so that you can get a "quick and dirty" measurement of the overall processing capabilities of your Azure landing zone. If the comparison shows big differences in performance and elapsed time between the before-migration database and the after-migration database, then you know that you have a match (or mismatch) in the expected balance of cost versus performance.

## CHAPTER 12

Postmigration Tasks

One of the best compliments I've ever given a direct report on her review was "Finishes projects with the same level of energy as when it was started." All too often, projects tend to trickle to a close once the big, headline-grabbing items are complete. Victory is declared too soon because the definition of "done" may not have been well defined. New projects that have been waiting to kick off take our attention, tasks at the end of a project are often mundane, and the boss has already sent out the kudos email congratulating the team on their accomplishment. Also, people would like to see their loved ones again after spending countless hours planning and executing a data platform migration. Whatever the reason, closing a project is rarely done with the same level of vigor as when it began, and the results can be expensive as legacy systems are not decommissioned or still running.

The tasks performed at the end of a project often have as much impact on the success of the project as those performed at the beginning. Failing to properly end a project can result in unwanted consequences later. Failing to update monitoring thresholds and policies may result in false-positive alerts or missed outages. Updating relevant documents and artifacts may be low priority – until a customer calls to report an issue and the engineer does not have the proper documentation to resolve the issue.

The following diagram details a standard migration project life cycle. It is not uncommon to see projects run out of energy and end prematurely sometime between the Validation and Cleanup/Optimization phases that you see illustrated in Figure 12-1.

© Kevin Kline, Denis McDowell, Dustin Dorsey, Matt Gordon 2022
K. Kline et al., *Pro Database Migration to Azure*, https://doi.org/10.1007/978-1-4842-8230-4_12

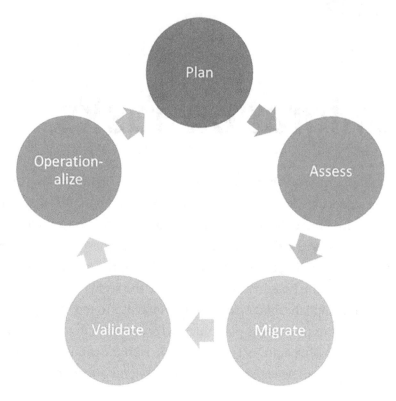

Figure 12-1. *A typical migration project life cycle*

In this chapter, we will review important considerations to ensure the success of your data platform migration. The concepts will be divided into technical and process-related items.

Decommissioning Legacy Resources

The process for decommissioning old infrastructure and machines may vary between organizations. Corporate policy, security and compliance, and regulatory considerations are factors in how and when legacy systems are deprovisioned.

It is generally best practice to take a phased approach to removing old systems following a migration to Microsoft Azure. A proposed process may look similar to the steps in Table 12-1 (the steps in the table are high level and illustrative only).

Table 12-1. *Steps to removing old systems*

Step Number	Action
1	Build cloud landing zone
2	Migrate pilot app(s) and source database(s)
3	Performance and UAT testing
4	Sign off on testing results
5	Migrate production apps and data sources
6	Cutover (unless running source and target systems in parallel)
7	Shut down source servers
8	Monitor application for 30 days
9	Back up source systems
10	Deprovision and remove source systems

Steps 7–10 are the postcutover tasks. Once data has been migrated and client traffic routed to the new environment, the source servers should be stopped or shut down. This step guarantees that no client requests are routed to the legacy servers. It is not uncommon for connection strings to be missed or to use IP addresses rather than Fully Qualified Domain Name (FQDN), in which case some requests may still route to the existing servers. Once the servers have been stopped, monitor closely to ensure no errors are received by client applications.

Stopping or shutting down the servers provides the ability to bring them back online quickly in the event of errors or failure. It is important to be able to get back to your starting point until the success of the migration is validated.

How long an organization waits before removing source systems is dependent upon multiple factors. Once the source systems are backed up, the decision to decommission them depends largely on the Recovery Time Objective defined for the source systems. If needed, the amount of time to restore a backup to a system may be a determining factor into whether to keep the source system "cold" for a period of time. Additionally, it is best practice to retain (or create) DNS records for the old hostnames to ensure any references to legacy names in code or connection strings will not break connectivity for dependent applications.

Validation and Optimization

Experience has shown us it is likely that not all issues will be detected in testing prior to migration. It may be an outlier use case or a difference between the source and destination data platforms. You may have that one legacy customer still using that old stored procedure you forgot about. Whatever the reason, it is critical that you validate your testing results with your results once the workload has been moved. In doing so, you will ensure that your cloud data platform continues to support your customers' business applications now and in the future.

Performance Issues

Some of the most frequent issues following a migration are performance related. Even after rigorous testing and the implementation of recommendations made by the Database Migration Assistant, some performance changes are to be expected.

I once moved a financial services client's databases from SQL Server 2014 to Microsoft Azure SQL Database. The move was pretty straightforward, and the only minimal updates were required (e.g., update legacy data types, add fully qualified object names in join code). The source hardware was eight years old, and the customer had not purchased extended support.

We provisioned the target environment with ample headroom when sizing the target databases. The migration and cutover went smoothly, and we celebrated our victory. A few days later, we began receiving complaints that front-office analysts – the people making actual financial transactions – were experiencing longer than usual runtimes on some key reports along with excessive blocking on the database instance. For this client, a delay in reporting could result in tens or hundreds of thousands of dollars in losses.

After extensive diagnostics, we determined that the report query performance was impacted by data skew. Because of parameter sniffing, a single plan is used for all parameters, and some parameters will perform suboptimally when data is skewed. A query hint was added to disable parameter sniffing, while the offending code was rewritten to use nested procedures using conditional logic. While this report had been tested prior to migration, we had not been able to replicate the exact production use case, which resulted in performance issues.

It is important to understand that even the most rigorous premigration testing regimens do not often catch every potential use case and performance issue. It is not uncommon for edge use cases to arise post migration, which were not accounted for in testing and result in complaints about the performance of an application. This story serves to illustrate that testing does not eliminate the need for close monitoring of critical metrics and systems post migration. In fact, it is always best practice to monitor the performance of your systems even more closely following a migration or other major change. This is when the 360-degree observability discussed earlier becomes important because your data platform may exhibit new behaviors not previously seen.

Business Impact Analysis

One common method we use to determine the appropriate BCDR solution is the Business Impact Analysis (BIA). The goal of the BIA is to determine RTO and RPO requirements based on how an outage of a system would affect a business based upon financial, legal, and reputation risk. The goal of the BIA is to create a requirements-driven, quantifiable determination of how an application should be protected. From there, the appropriate solution can be determined.

As an example, consider a financial service business with the line of business data applications shown in Table 12-2.

***Table 12-2.** Line of business applications*

Application	Data Platform
Trading Platform	Microsoft SQL Server 2016 Enterprise
Reporting	Azure SQL Database
Files	Blob Storage
Customer Service	Microsoft SQL Server 2016 Enterprise

In conducting a Business Impact Analysis of three line-of-business applications, the determinations shown in Table 12-3 are made.

Table 12-3. *Business Impact Analysis requirements*

Application	Risk: Financial	Risk: Legal	Risk: Reputation	RTO	RPO
Trading Platform	High (>250k)	High	High	<5 min	<30 sec
Reporting	Medium (100k–250k)	Low	Medium	<2 hours	<15 min
Files	Low (<50k)	Medium	Low	<4 hours	<4 hours
Customer Service	High (>250K)	Low	High	<2 hours	<4 hours

Note   An additional value, **Maximum Acceptable Outage (MAO)**, is often included in the BIA. The MAO defines the amount of time a system can be down before the viability of the business is threatened. I have decided not to include it here for the sake of clarity.

Once the requirements have been defined, a tier is assigned by service-level agreement (SLA) using a map similar to the one shown in Table 12-4. (The values used in Table 12-4 are illustrative and will need to be modified for your particular business case.)

Table 12-4. *BCDR tier by SLA*

Recovery Time Objective (RTO)	Recovery Point Objective (RPO)			
	SLA	<15 minutes	<2 hours	<4 hours	<8 hours
	<15 minutes	Tier 1	Tier 1	Tier 1	Tier 2
	<2 hours	Tier 1	Tier 2	Tier 2	Tier 2
	<4 hours	Tier 2	Tier 2	Tier 2	Tier 3
	<8 hours	Tier 2	Tier 2	Tier 3	Tier 3

Using the values from Table 12-4, we can assign the applications to tiers as shown in Table 12-5.

Table 12-5. *Application BCDR tier assignments*

Application	BCDR Tier
Trading Platform	1
Reporting	1
Files	3
Customer Service	2

The appropriate BCDR solutions will be applied to meet or exceed the SLA defined at each tier. For example, the Trading Platform application is a Tier 1 application running on Microsoft SQL Server 2016 IaaS virtual machines in Azure. The low RTO and RPO requirements dictate a transactional solution such as Always On Transactional Replication (although the DNS change required by the latter may introduce unacceptable downtime while it propagates). Alternatively, the Customer Service application can tolerate longer RTO and RPO as well as expensive solutions such as Transaction Log Shipping. Finally, as a Tier 3 application, the Files system can leverage Blob storage with backup/restore.

If I sound passionate about this topic, it is because I am. In my job as Principal Cloud Architect supporting the US Army's Enterprise Cloud Management Office (ECMO), we support applications where downtime is measured in the number of lives lost. Given those stakes and the scale of the applications we support, it is critical that we have a reliable, quantifiable method to determine how we keep applications online. However, we also support far more mundane applications that do not have the same criticality. The goal of this exercise is to define the appropriate BCDR solution based upon the application requirements. Substantial reductions in cost, complexity, and business risk are realized by not using overly complex and expensive BCDR solutions for applications that do not require them. This is a common way organizations overspend on Azure, and applying a methodology such as the BIA can have a dramatic impact.

Right-Sizing Your Data Platform

In an earlier chapter, we discussed how the planning paradigm for on-premises systems and Microsoft Azure are different. In short, when purchasing hardware, we intentionally purchased unused capacity as headroom for growth. In Microsoft Azure, you can dynamically scale your platform to accommodate your workload. Overprovisioning resources is a key driver of expenses, and we should run "hotter" in Azure. After migration, it is very important to monitor your systems to ensure they are sized properly – not over- or underprovisioned. Overprovisioned means your organization is paying for unused capacity. Underprovisioned means your systems will not support the business workload. Monitoring capacity is an ongoing process that should be done on a regular basis with well-defined thresholds and rules for sizing. However, it is even more critical following a migration as you validate the results from your testing.

Some organizations intentionally oversize target resources for the initial phase of a migration to ensure the systems meet performance requirements and then right-size them after validation and tuning. This approach is a little more expensive at the outset but makes sense if there is uncertainty about the workload there are gaps in performance testing. Just be sure to go back and right-size the resources to avoid unnecessary expenses for unused capacity.

In Chapter 5, we talked about the importance of establishing effective baselines and meaningful thresholds for alerting and notification. As with BCDR, it is important to have predictable and quantifiable measures to determine the health and performance of an application or system. Some performance profiles may change after migration due to changes in underlying compute or storage performance, SQL Server Engine version changes may result in suboptimal execution plans, or connectivity and routing between on-premises and Azure resources may cause unforeseen latency, which did not surface in testing.

Technical Debt

Technical debt occurs when organizations implement workarounds, temporary solutions, or take other shortcuts that do not consider the long-term effect on the long-term consequences when developing or implementing solutions. Examples of tech debt include the following:

- Implementing temporary solutions to solve persistent issues within a system

- Relying on legacy systems or platforms that interfere with rapid deployment

- Delaying upgrades on infrastructure resulting in increased costs to maintain the systems and preventing developers from implementing new features

Technical debt may be incurred inadvertently or on purpose as a strategic decision. However, the net result is an increasing drag on innovation and production. Over time, the resources required to pay down technical debt create increased drag on productivity as more resources are diverted to the technical debt backlog. Even if no new tech debt is incurred, existing tech debt will exert an increased effect on an organization's ability to innovate due to reliance on legacy systems and ad hoc, temporary solutions. Moreover, the cumulative expense associated with increasing technical debt was estimated in 2015 by Gartner at over $1 trillion globally and over $200 million on average for large, publicly traded companies (`www2.deloitte.com/content/dam/Deloitte/xe/Documents/technology/me-consulting_technical-debt.pdf`).

Migrating to the cloud can help organizations avoid or retire technical debt in a number of ways. First, offloading the responsibility for upgrading and maintaining underlying infrastructure frees human and capital resources to focus on adding value to the business through feature development and increased efficiency. Finally, the cloud consumption model creates a predictable model as expenses are shifted from capital to operational expenditures.

One of the most common sources of tech debt is inconsistent configurations across environments. These "snowflake" environments often manifest as inconsistencies in Dev-Test-Prod environments or as differences between business units or geographic locations. In the former, the result is unpredictability in how code will perform when deployed to prod as the results from testing cannot be trusted to accurately reflect the production environment. In the latter, administrators must have different policies and procedures for each environment due to differences.

Some of these issues and inconsistencies can be mitigated prior to migration. However, Microsoft Azure allows organizations to leverage robust orchestration, automation, and event-driven pipelines to ensure configurations are consistent between environments. By converting infrastructure to code, organizations can leverage existing CI/CD pipelines and native Azure technologies to guarantee consistency in technology and processes across environments and geographies.

Summary

Moving data workloads to Microsoft Azure requires a great deal of planning and assessment prior to the actual migration. But the work does not stop there. Microsoft Azure provides the data professional with tools and capabilities that were not previously available, but they require careful attention to the postmigration tasks and processes in order to be successful. The sheer increase in options and methods alone can be transformative for how the Microsoft Data Professional does their job, but without migration follow-up, those same options can result in inconsistencies and issues that negate many of the efficiencies that drove the migration in the first place. Simply put, getting the workload to Azure is one thing; running it there is another.

CHAPTER 13

Post Mortem

The last remaining item in your data migration to Azure is the post mortem. The post mortem is a process that occurs at the end of a project to look at everything from start to end to review what went well, what did not, and what can be improved for future projects. Winston Churchill once famously said "Those who fail to learn from history are doomed to repeat it," and this is true in everything we do including technology. Taking time to discuss and learn from the previous project can increase efficiencies and productivity via actionable steps going forward.

If you are familiar with or have worked for a company that practices such methodologies as Scrum or Agile, then this is probably already a familiar concept. This is also commonly referred to as a retrospective within those environments. The difference is that retrospectives may happen more frequently. And they are generally performed at the end of a sprint and focus on reviewing items from the past Sprint. Post-mortem functions the same way but occurs at the end of a project and is meant to look at everything as a whole. Retrospectives throughout the project are highly valuable and would be a great asset to your team, but at a minimum, you will want to review during the post-mortem process.

I have worked for companies before where this stage of the process was either skipped completely due to its perceived low importance or done in quick fashion just to mark a checkbox. This can be a meaningless activity that helps no one, or this could be a catalyst for organizational and team changes for the better. No project is an unmitigated success or failure, and there are always lessons to be learned.

Note No project is an unmitigated success or failure. There are always lessons to be learned.

© Kevin Kline, Denis McDowell, Dustin Dorsey, Matt Gordon 2022
K. Kline et al., *Pro Database Migration to Azure*, https://doi.org/10.1007/978-1-4842-8230-4_13

In this chapter, we are going to focus on how to successfully run a post-mortem process so that it is meaningful and can help you go forward. One of the most essential components is the meeting that will occur that will include the team members to talk through items; however, it's more than just that, and we will review what.

The Benefits

With the post-mortem process, you are only going to get out of it what you put into it. If no one takes it seriously, then it's probably not going to be very beneficial for anyone. You need the team to be engaged with this to really see its benefits. This starts with someone being an advocate for this part of the process, which might be you the reader. To advocate for this and sell to your team, then it's important to understand the benefits of what this provides.

First, this part of **the process is intended to help those that have been performing the work**, and it's a chance for their voice to be heard. In the current working landscape, one of the top reasons people leave companies is because they say they don't feel valued or appreciated. We have experienced this throughout our career and even left jobs as a result. Additionally, we have had several colleagues that have done the same. As people, we want to see that our hard work and expertise are appreciated and acknowledged not just through words, but also through actions. A manager can shower you with verbal praise, but if you are left out of key decisions or your input is ignored, then it does the same thing.

There are unquestionably bad and selfish managers that exist out in the workplace, but sometimes, managers' intentions can be misrepresented. Our days, like many others, are really busy and usually consist of a checklist of things that we need to get done in too little time while spending our entire day in meetings. Taking time to acknowledge, appreciate, and listen to our staff is something that we have to make time for; otherwise, it's easy to overlook. It's not that we don't appreciate them otherwise, but we need to let them know. Processes such as retrospectives and post mortems provide us with built-in opportunities to show this appreciation to our teams.

Note   It is important to make time to acknowledge, appreciate, and listen to your staff; otherwise, it's easy to overlook until it's too late.

The next benefit is that it **improves efficiency.** One of the key components of a post-mortem process is to uncover things about the project that didn't go so well and determine ways to improve it going forward. This can result in process changes to your team but can also result in organizational changes as well. We will cover this in greater detail throughout the chapter.

This process can also **improve the morale of the team**. As mentioned earlier, it provides an opportunity for the team to have their voice heard and influence change. The post mortem is also an opportunity to celebrate the wins and success of the project. Both of these are ways we can acknowledge and show appreciation to them, which increases overall morale. Talking through problems and celebrating the successes will bring the team closer together and help build excitement for the next project.

This should also open the doors for **better collaboration going forward**. Discussing issues and successes is a great way to understand how people think and work, and we can use that to learn from them. By creating opportunities to allow us to understand others' perspectives on the team, it causes us to be more empathetic toward them in future collaboration. Your team is likely going to be made up of a lot of different personalities that are going to respond differently in different situations; learning what works best for them can help bring out their very best.

The last benefit is that it is an opportunity to **provide closure on the project**. It is a deliberate action to celebrate the success, reflect back on what you accomplished, and look for ways to improve moving forward. Without it, these things rarely happen with any consistency or structure.

Another thing we like about providing closure is that it signifies the end of the project. Have you ever been part of a project that drags on and on; despite you feeling like it has been completed for weeks, it stays open? The post-mortem process brings a clear-cut end to it and helps signify a separation between the existing project work, operational support, or the next project.

While there are probably more advantages than what is listed here, hopefully, you can see the advantage of this process and be an advocate of its importance to your team.

The Process

If you have participated in a post mortem or retrospective in the past, you may have viewed this process as just a meeting. While one of the primary components is a meeting, it actually entails a lot more than just that. After all, you are asking your team to reflect back on their experience and provide info back. If you only have a meeting with no thought given to the process, then the process is going to be less effective. The only feedback you are going to get is whatever comes to the participants' minds at that moment.

The first part of the process and arguably the most important is having an owner of the process. Someone has to take the reins of owning this process from start to end. Generally speaking, this is handled by either the project lead or the project manager. While the post-mortem process is successful based on the effort from the entire team, this person is the one responsible for making sure each task gets completed. This includes handling coordination, scheduling, documentation, and delegation of activities. They are the glue that binds everything together and makes sure the process works. It does not mean they are responsible for every component, and it is common to have parts of this process delegated out to other members of the team.

After an owner is established, it's time to get to work. There is pre- and postwork that needs to occur to make this part of the process successful. Here is an overview of some of the important components that we will discuss over the coming sections:

- Post-mortem questionnaire sent out ahead of time

- Electing a moderator

- Schedule the workshop

- Structure for the workshop

- Meeting rules and guidelines

- The takeaways

Another thing to note here is that while this is the last step in the process, you shouldn't be waiting until the end of the project to start thinking about it. The post-mortem process should be something that you factor into the project during the initial project planning and accounted for in the overall timing. Just like timing being factored for scoping and development work, you should be factoring the post-mortem.

The last thing we want to mention about the process is that you are only going to get out of it what you put into it. If no one takes it seriously, then it's probably not going to be very beneficial for anyone. You need the team to be engaged with this to see its benefits. This part of the process is intended to help those that have been performing the work, and it's a chance for their voice to be heard.

Note   The success of the post-mortem process is determined by the effort put into it. If no one takes it seriously, it's not going to be very beneficial.

Post-Mortem Questionnaire

Before the post-mortem meeting, you need to put some thought into specific questions you would like to have answered to generate good conversations. The purpose of the meeting is to discuss what went well, what didn't, and what can be improved on for the future, but if this is all you tell your team, then you are probably going to get some quick answers without a lot of depth. You will want to generate subjective questions that force people to think beyond what's just at the surface.

Once you develop a list of questions, send them out to the team well ahead of the meeting. We recommend at least one week prior, but ideally with as much time as possible. This way people have an opportunity to put some thought into their answers rather than having to rush through it. Remember, this shouldn't just be a check box, but something that is meaningful so you want to provide adequate time.

The team members you will want to send this to are the same ones that will be participating in the meeting. These are people who played meaningful roles in the project, not necessarily everyone who performed a task associated with it. For example, with an Azure migration, you will absolutely want to include the folks who performed the actual migration, but you may not want to include the team that only created service accounts for the project. As a leader, it will be up to you to decide who may be able to provide meaningful insights into the project and who may not.

Additionally, with large projects, you can have a lot of people involved, and it's possible to have too many people part of the post mortem where it can become a jumbled mess. We recommend keeping the meeting to under 15 people; however, if you have a lot more than this involved, then you may need to consider limiting the attendees. One option is limit the discussion to the leads or senior-level resources to keep attendees down. As you can imagine with a large group, it can be difficult to navigate the topics with everyone having varying opinions, so recommend keeping it smaller so it stays meaningful. The other team members still deserve to have a voice though, and we still recommend including them as part of the questionnaire that gets sent out in advance.

You can get very detailed with the questions you ask, and it is really up to you to build this list. There are numerous online resources and books available out there that cover this subject in much greater detail that can help you question lists. We do recommend starting with fewer questions initially if this is something that the team is not used to. You don't want the team getting overwhelmed with looking at the list or trying to rush through it because there is so much to do. Quality is much more important than quantity at this point. As the team grows though and everyone begins seeing the value in this, then you can start adding to it. Additionally, avoid questions that are black and white or that have a right answer; you want your questions to be subjective.

Don't worry about trying to create the perfect question list, you will drive yourself crazy. Just get started and you can tweak as you get more and more experience with this process. You will learn and grow and be able to refine the list to what's meaningful for your team. To give you some ideas though, here are some of our favorite questions to ask:

- Are you proud of the outcome of the project? Please explain why.

- How would you rate the overall project plan and execution of it?

- What was the most frustrating thing about the project for you?

- How would you have handled that differently or suggest handling it differently in the future?

- Did you feel you had a voice in the project and it was heard?

- What was the best part of the project for you?

- What methods or processes do you think worked well?

- What methods or processes did you find challenging or frustrating to use?

- If you could change anything about the project, what would you change?

- Do you feel the right people and teams were involved and engaged in this project? Please explain.

- Did you have adequate support from all pertinent parties including leadership, other teams, customers, business sponsor, etc.? If not, then in what ways do you think we could improve their participation?

- Did you feel you had adequate time and resources to perform the duties assigned to you?

You will notice that the questions are very subjective and ask for more than just a yes or no answer. Avoid using objective questions.

Note   Use subjective rather than objective questions.

Once the questionnaire is sent out, request that the team answer the questions and send it back to you or whoever the owner is. You may also consider having the responses sent back to a third party to compile for the owner so you can guarantee anonymity. Once sent, set a deadline for the responses and make sure there is enough time for the owner or recipient to read through and aggregate the results. This is good practice with any size group, but it's especially important when not everyone is able to attend or when you have quieter and shy people on the team. During meetings, some people have a tendency to be overshadowed by the more outgoing and vocal members of the team, and this helps with that.

Once the results are returned and you have feedback, the owner can usually quickly identify common themes in the things that went well and didn't. Make a list of the top three to five things in each and add them to the agenda to make sure they are discussed during the post-mortem meeting. For example, if 75% of the team stated that they spent too much time meeting about the project rather than doing work (a common item), then it's a good indication that it is a problem area to the team that warrants discussion. In addition to identifying the issues, look at the proposed solutions and list them out with it. We will review more details about the collective meeting to discuss these items later in the chapter, but we do want to highlight that there is not a perfect solution to every problem and there could be common complaints that it isn't a better or doable solution for. Common issues we see that don't get resolved in the corporate world are ones that need to be solved by money. Businesses have to weigh the risk/cost benefits, and it doesn't always fall in the engineers favor.

I had a manager once before that would always tell me "don't bring me problems, bring me solutions" when I would approach him with an issue. If I did not have a solution to propose, he would not listen to me. It completely changed my way of thinking and has honestly helped me become a better leader and technical professional. Since then, it's been something I have tried to instill in my daily life. This is the same concept you want your staff to have when completing this questionnaire. It's not an opportunity to complain, but an opportunity to offer meaningful feedback on how things can be improved. If you're only complaining about issues and not offering feedback on how to fix it, how can you expect others to do it?

When my manager would tell me this, it was not because he always expected me to have the right or final answer, but it was to make me stop and think about it. On several occasions, I took solutions to him that got shot down for various reasons (usually financially related), and we went with what I considered a less optimal solution, but I always had a good understanding of why.

Note   If someone is going to shed light on a problem, they need to also have a proposed solution of how they would fix it.

Elect a Moderator

Prior to the post-mortem meeting, we highly recommend electing a moderator for the meeting that has not been associated with the project to help facilitate things. The owner of the post-mortem process can still lead the conversation, but because they have been involved with the project, they probably have their own views on things that could differ from the group. The moderator is responsible for keeping things focused and provides balance and order to what can sometimes become heated conversation.

It helps as well that the moderator carries the respect of the team either by personality or by rank so people listen when they talk. The moderator will be expected to take charge when warranted to keep the conversation professional and on topic. But you do not want someone that would inhibit the team from wanting to share for fear of retribution. This is usually why we prefer someone from a different team under a different reporting structure to help with this. Their role is not to influence the conversation, only to maintain order.

The moderator's primary role is to keep things professional and on topic, but they can also be utilized to help take notes, capture the feedback, or be part of later conversations with leadership to impact change. The post-mortem process owner would need to coordinate with them to determine if there are additional activities they would like them to perform.

In the event you do not have a good candidate for a moderator or do not feel like this is necessary, then you can look at other options. We do not recommend having a single person from the project handle this though, but you can make moderation a team activity. This means every person who is part of the process plays a role in ensuring order and professionalism. The reason we don't recommend having a single person do this is because they are not unbiased and could be the one causing the issues.

Post-Mortem Workshop

At this stage, you have sent out the questionnaire and designated a moderator, and you are ready to plan the post-mortem workshop. Hopefully, you had adhered to earlier advice and made this part of your overall project plan, but if not, then it's fine; you may just have some additional challenges to work through.

The first thing to do is consider how much time you need and find a time to schedule if you haven't already. The amount of time can depend on several things such as the number of attendees, personalities of the team, size of the project, level of feedback from the questionnaire, overall success of the project, and so on, so it's going to be subjective. Remember that this is designed to help the team so you want to provide adequate time. We usually recommend at least 90 minutes though as a starting point. Much shorter than this, then you risk not being able to cover everything, and much longer than this, then people may start to lose interest.

Scheduling may not be as big of an issue if you factored into the overall project, because time should be held for it well in advance. If not, you will need to find some time to schedule. If you have a good project manager, then they are a whiz at finding time in people's schedules.

Since it's the final step in the project though and usually a good time to celebrate, we like facilitating these around lunch and ordering in lunch for the team. Something about filling our bellies with food together that adds an additional layer of camaraderie. Also, if you had not scheduled the meeting far enough in advance, lunchtime may be a good time to find openings in folks' schedules. If you do factor lunch into the meeting, then we do recommend providing at least an additional 30 minutes to allow people time to eat.

Once you have the meeting scheduled and sure you have the right group part of it, then you need to consider how you are going to spend the time and structure a clear agenda. By the time you start putting together the agenda, you should have feedback from the team on the questionnaires that were sent out previously. You can read through these and pull out some of the common themes to make sure these are discussed. It is going to be difficult to address every little concern; however, you definitely want to hit on the most prevalent. The one-off issues people report that are more personal than these can be handled in a different setting.

Note   Structure the meeting with a clear agenda and focus on the most prevalent items from the questionnaire.

The meeting can be scheduled to work in whatever way works best for you; however, we have provided a template that can assist you. From our experience, most of the time we have spent in the post-mortem meeting is spent on discussing things that can be improved on, so make sure you have plenty of time for that.

Example agenda:

- Review the ground rules – <5 minutes

- Quick recap of the project – 5 minutes

- The wins! What went well – 15 minutes

- Areas of improvement – 35 minutes

- Actionable steps going forward – 20 minutes

- Team appreciation (end on a high note) – 10 minutes

In our example agenda, the first thing you will want to start with is laying out the ground rules for the meeting. This is to make sure everyone understands what this is intended for and what it's not. In the next section, we will review some examples of what these are.

Next, have the project lead give a brief overview of the project and what was accomplished. Did you complete what you set out to do or not? This is a good opportunity to give thanks to the team for everyone's hard work to get it to this point.

The next part really starts diving into the project. We like to start with the wins and successes and talk through the things that went well. The project lead should have a list of the top items garnered from the questionnaire and can use those as starting talking points. The successes are important to note because it increases morale and helps the team see that their hard work is appreciated. Once you go through the most common things, open it up for discussion from the group. Set a time limit here to make sure you have time to cover other topics.

The next part is the part that gets interesting. It is looking at the things that may not have gone so well and how we can improve them. Once again, the project lead should have a list of the most common themes found in the questionnaire to start the discussion with. Go through each and garner feedback from the team. Try to make sure the conversation stays meaningful so it's not just everyone repeating the same complaint or attacking other individuals on the team. This is not an opportunity for people to complain about things that didn't go well, but an opportunity to look at how they can be improved on in the future. Take notes, discuss solutions and next steps, and move on to the next one. Once you get through the most common elements, open up to the team to discuss the most pressing items for them if they have not already been discussed. If it's a personal independent issue, then address that separately from this meeting.

Once the issues have been gone through, make sure you discuss actionable steps going forward. Talk does no one any good if it doesn't turn into meaningful action. With each actionable step, document it and assign an owner from the team that will be responsible for working through that item. The project lead is not responsible for every actionable step, and these should be divided up among the team. We will discuss this more later in the chapter.

Sometimes, during these conversations, things can become tense, so we always suggest trying to end the meeting on a positive note. Remember that despite the challenges that occurred, you persevered and completed the project, and that is to be celebrated. Everything we do is a learning opportunity to get better for the future, and with each subsequent project, we get a little better.

The last thing to mention is that you want at least one person taking really good notes that can be shared with the team afterward. You can utilize the moderator for this, have an elected person on the team, or even utilize multiple people, but you want to make sure it's captured and saved. In the notes, avoid using specific names of people and try to keep them general. We also do not recommend recording the meeting and saving it for this same reason.

Meeting Rules and Guidelines

Because one of the core components of this meeting is eliciting feedback from the team on what did not go well, clear guidelines need to be set about what is and is not acceptable. This is to make sure everyone stays on topic, stays professional, and feels comfortable sharing. Because the project lead is not unbiased and may have their own opinions, we recommend utilizing the moderator we previously discussed to handle enforcement of these guidelines. Or if that is not possible, then utilize the entire participating team to make sure these are adhered to.

You can create your own rules and guidelines, but here are some of the ones that we use with our teams. Let's start by looking at four simple rules we share with our teams before the post-mortem meeting.

Rule #1: Be professional and no personal attacks.

If you could listen to every unfiltered thing people say or think about you and their colleagues, none of us would ever get along. People get angry, offended, hurt, and so much more from those around them all the time, and sometimes, they blame the issues of a project on these people. And sometimes, those emotions can come out during this. The post-mortem meeting is not the place for that. The number one rule in any post-mortem meeting is that people remain professional and abstain from any personal attacks. The moderator should help control and limit this since sometimes, it can involve the project lead themselves. In instances where there are repeat personal attacks, then the moderator can at their discretion request that someone leave the meeting.

Rule #2: If you are going to bring up an issue, you need to propose a solution.

We looked at this earlier as well, but if someone is going to bring up an issue, then they need to have a proposed solution of how they would fix it. The post mortem is not a complaining session; it's an opportunity to identify issues and improve them going forward. Complaining without solutions is not acceptable.

Rule #3: Limit feedback to only items relevant with the recent project.

The next rule is to limit the feedback to only items that are relevant to the project being discussed. This is not an opportunity to discuss personal work issues or issues related to another project you are working (or have worked) on. For example, this is not a place to express your grievances that the vending machine doesn't have your favorite soda anymore. The moderator should help monitor the items brought to make sure they are in reference to the recent project.

Rule #4: Speak up!

Next, each member of the team should speak up to discuss any improvements they think that can be made. Over the years, we have had several complaints about an issue only for the person to remain silent when provided the opportunity to share. The post-mortem process is an opportunity to speak up; utilize it.

In addition to the rules, here are a few guidelines that we use when scheduling the meeting. These are not necessarily shared with the team but are important for the organizer to keep in mind while they are facilitating the post-mortem process.

Guideline #1: Limit attendance to those that performed work as part of the project.

As the organizer, you need to make sure that you are limiting the attendance to just those that performed work as part of the project. You may have other people or members of leadership that are interested in attending, but this is not the place for them. The information can be collected from the meeting and shared with leadership at a later time. By having them attend, you may have members on your team afraid to speak up, which is the opposite of what you want. Everyone in the meeting (outside of the moderator) should be able to add to the conversation.

Guideline #2: Create an open space where people feel comfortable sharing.

It is important that the team members that are part of the post-mortem process feel open about sharing their thoughts. Oftentimes, people hold back out of fear of retribution or being viewed differently because they worry about the impact it can potentially have on raises, promotions, or their overall career. This is not something that we think there is a magic solution to overcome because so much of this depends on the culture created around them. A culture that fosters openness will probably generate more responses, whereas cultures where leadership makes decisions independently of what their employees think may not. You can though limit the attendance to exclude leadership not part of the hands-on work, and you can keep information shared anonymous when having later conversations. In a lot of cases, this is a time thing, and people have to feel comfortable opening up first. Hopefully, you have some people (like myself) who do not mind being outspoken who can speak up for some of the common issues.

Actionable Change

During the post-mortem meeting, there is going to be a lot of conversations and a lot of information to digest. It is very important that someone be taking really good notes throughout this meeting so nothing discussed gets lost. Unless delegated to someone else, the responsibility defaults to the person who owns this process. Following the meeting, notes about the discussed topics should be compiled and sent out to the team for reference.

One of the most important outcomes of the meeting is the list of actionable steps that is generated during the meeting. A post mortem that does not impact future action is a waste of time. This list of items is compiled by the group as whole and captured via the note taker. These are actionable steps such as process changes that the group agrees should be implemented going forward.

Note   A post mortem that does not impact future change is a waste of time.

You need to have a plan of how to address these and who is owning it. As mentioned before, this does not all fall on the project owner and can be assigned to anyone on the team that is capable. Set deadlines for when the item should be completed and hold them accountable the same way you would with other tasks.

Some items the group can implement on their own accord. For example, the group agrees that managing tasks via Excel is not a great solution and the team wants to begin using Azure DevOps for tracking. This is a change that the team can probably make without needing to go through leadership approval and is just a process change for the team. Other items may require additional leadership conversation. An example here would be if the team wants to purchase a new development tool to increase the efficiency of the team, but it comes with a significant investment. The team may agree it's the right solution, but you still need approval from leadership to make it happen. In addition, sometimes, there are cultural or organizational proposed changes that cannot occur at the engineer level that require leadership involvement.

If you have engaged leadership, then they are probably going to be interested in hearing about some of the most prevalent items from the post-mortem process and how you intend to improve on them. If so, then the conversation to share these details would be a good opportunity to raise issues requiring their involvement to their attention. For less engaged leadership, you may need to reach out to them to bring some of the issues to their attention. If you have ever worked with leadership to bring about change, you probably know this is not usually a one-and-done conversation and often will require more work on your end. Whether it's providing justification or taking a leading role in the implementation of the change, there is probably going to be more work ahead. However, if the team agrees it's important to them, then it's worth the investment to do what you can to make it happen.

One thing we will add is avoid naming specific individuals when talking with leadership and keep feedback general. While it may seem harmless to you to share a specific person's thoughts, it can sometimes be taken differently than you intended. And if it gets back to the person who said it, then it damages the trust they have in you and potentially causes them to hold back helpful insights going forward. Think of the insights gathered from the post-mortem as from the collective team rather than an individual and always communicate it that way.

The Other Items

It is very unlikely you will be able to cover all of the answers to the questionnaire within the post-mortem meeting because there would never be enough time. The goal of the meeting is to hit on the most prevalent items first and then, as time permits, look at some of the lesser noted items. However, that doesn't mean these things aren't important and they should still be considered.

Oftentimes, what we see with the one-off items is that they deal with something more personal in nature to a person that doesn't necessarily resonate with the rest of the team. These are things that probably need to be discussed in a more private setting rather than with the entire team. Some of it could be silly items, but some could reflect on more serious issues.

Here are some examples of things you could see:

- I was treated differently based on race, age, sex, ethnicity, or sexual orientation.

- I did not like working with a particular employee.

- I was managed differently than other people on the team.

- I had to carry more weight in the project than other team members.

- I was never able to speak during meetings because another team member would never stop talking.

- I only have one computer monitor and other members of the team have two.

- My laptop is old and slow.

- The person sitting next to me never wears deodorant.

As you can see, these range in their version of seriousness. Some of these can be handled with just having some short conversations or email with the individual, but some may require a lot more attention or even escalation to other people. While this book is not a guide on how to handle every issue that may come up during the post mortem, it is important that you don't just ignore it. If it was important enough for someone to write down or share, it's important enough to address.

Note   If an issue was important enough for someone to write down or share, then its important enough to address.

Following the post-mortem meeting, look back at some of the one-off issues that may not have been discussed to see if this is something that you can just address directly or if it warrants additional conversations and take action on it. We are a huge advocate for creating the best work environment as we can for our staff, and one way we can do this is by listening and responding to their issues (even on the silly stuff).

Show Appreciation

A lot of the attention of the post-mortem process is focused on the issues that can be fixed, but do not neglect the need to show appreciation to the team. For projects such as a new Azure migration, there are a lot of complexities and new technology being introduced to the team that will stretch and grow them. Be sure that you recognize and show appreciation for the effort and diligence they put into it. They may not seem like they care, but we can assure you they do, and it could mean the difference between them taking another role they are being courted for or not. Personally, when we feel our team and leadership appreciate the work we are doing, it makes it much harder to even consider other opportunities.

Note   Be sure to recognize and show appreciation to the team for their hard work and diligence throughout the project.

In Chapter 7, we looked at ways to celebrate wins, and this is a perfect opportunity to apply those. If you have a budget, scheduling a dinner or activity is usually appreciated, but if not, there are still other things that you can do. To recap, here are some different things you can do to celebrate the completion of the project:

- Handwritten note

- Lunch or private meeting arranged solely to acknowledge a team or person

- Acknowledging the person or team in a company-wide communication

- A gift (if you have the means to do so)

- Day off or letting them off early

- Monetary items such as a promotion, raise, or bonus

- More responsibility or utilizing for special projects

- Team outing

Just whatever you do, make sure that the team realizes how much you appreciate them and all the hard work they put into this.

Onto the Next Project

You have done it and you are at the end of the project. You have popped the champagne, clinked the glasses, and celebrated all night long. So what's next? Well, it's now time to move to the next project and start all over. Hopefully, as you have journeyed through this project, you have gained important knowledge and become a better professional than when you started. Hopefully, you have learned valuable lessons that you can take into the next project to make it a better experience than the last.

Summary

In this concluding chapter, we discussed the importance of taking time to reflect back during the post-mortem process. This includes looking at what went well, what did not, and what can be improved on for future projects. If we do not take the time to learn from past mistakes, then we are going to be doomed to repeat them. And if we do not take time to celebrate wins, then the staff may not feel their hard work is appreciated.

While there is both pre- and postwork that needs to occur, the core component of this is the post-mortem meeting. This meeting includes the staff that worked on the project and is founded on their subjective views about successes and shortcomings of the project. The staff should be presented with a list of questions ahead of the meeting that they think through answers on to discuss during the meeting. The goal of the meeting is both to celebrate and also create actionable steps for meaningful change going forward that improves the experience for the next project.

Index

A

Access control, 205

Action groups, 122, 124, 125

Active Directory (AD), 124, 205, 217, 219, 220, 222

Actual costs, 117

Advanced data security
    data discovery and classification, 239
    threat detection, 238, 239
    vulnerability assessment, 240

Always encrypted, 232, 236–238

AlwaysOn availability groups, 54, 85, 279, 280

Amazon Web Services (AWS), 127

Analysis and testing phase, 37

Anomalies, 35, 117, 131, 145, 238

Anomaly-based alerting, 146

Application Performance Management (APM) tools, 133

Application Platform Monitoring (APM) tools, 149

Artificial intelligence (AI), 148

Auditing, 40, 41, 109, 202, 221, 229, 240

Authentication, Azure SQL
    administrator accounts, 217–219
    configure and utilize AAD authentication, 222, 223
    connection tips and tricks, 224, 225
    users/groups/practices, 220–222

Auto-failover groups, 58

Automated testing, 36, 53, 288

Azure Active Directory (AAD), 3, 205, 217, 222

Azure administrator roles, 218

Azure advisor, 112, 125–127, 132

Azure billing entity hierarchy, 111, 112

Azure blob storage, 2, 260, 278, 279

Azure calculator, 83, 84, 88

Azure container instances, 2

Azure core services, 3

Azure Cosmos DB, 2, 23, 209, 254, 260

Azure Database for MySQL, 2, 264

Azure Database for PostgreSQL, 2

Azure data factory (ADF), 279

Azure disk storage, 2

Azure ExpressRoute, 2, 268

Azure file storage, 2

Azure Hybrid Benefit (AHB), 81, 87–91

Azure kubernetes service, 2, 23

Azure load balancer, 2, 54

Azure migration, 186, 288, 309, 320

Azure monitor, 148–150

Azure networking services, 2

Azure Portal, 2, 4, 15, 96, 103

Azure Pricing Calculator, 79–83

Azure Private Link, 209, 210, 216, 229

Azure Purview, 242, 258–261

Azure region, 5, 38, 55, 58, 214

Azure reservations, 95–98

Azure resource groups, 4

Azure resource hierarchy, 111

Azure Resource Manager (ARM), 4, 10

© Kevin Kline, Denis McDowell, Dustin Dorsey, Matt Gordon 2022
K. Kline et al., *Pro Database Migration to Azure*, https://doi.org/10.1007/978-1-4842-8230-4

Printed in the United States
by Baker & Taylor Publisher Services